Innovations in High-Speed Railway Engineering - Aerodynamics, Infrastructure Resilience, and Sustainable Mobility

Edited by Masoud Mohebbi and Rasul Mohebbi

Published in London, United Kingdom

Innovations in High-Speed Railway Engineering - Aerodynamics, Infrastructure Resilience, and Sustainable Mobility
http://dx.doi.org/10.5772/intechopen.1009726
Edited by Masoud Mohebbi and Rasul Mohebbi

Contributors
Anjali Awasthi, Dongqin Zhang, Feng Haonan, Guo Hongli, Ji Yingjie, Liang Zhiguo, Masoud Mohebbi, Ma Xiaojiao, Rasul Mohebbi, Yong Lin Ren

First published in London, United Kingdom, 2026 by IntechOpen
IntechOpen is the global imprint of INTECHOPEN LIMITED, registered in England and Wales, registration number: 11086078, 167-169 Great Portland Street, London, W1W 5PF, United Kingdom

For EU product safety concerns: IN TECH d.o.o., Prolaz Marije Krucifikse Kozulić 3, 51000 Rijeka, Croatia, info@intechopen.com or visit our website at intechopen.com.

British Library Cataloguing-in-Publication Data
A catalogue record for this book is available from the British Library

Innovations in High-Speed Railway Engineering - Aerodynamics, Infrastructure Resilience, and Sustainable Mobility
Edited by Masoud Mohebbi and Rasul Mohebbi
p. cm.

This title is part of the Civil Engineering Book Series, Volume 20
Topic: Transportation Engineering
Series Editor: Assed Haddad
Topic Editor: Saúl Antonio Obregón Biosca

Print ISBN 978-1-83635-875-6
Online ISBN 978-1-83635-874-9
eBook (PDF) ISBN 978-1-83635-876-3
ISSN 3029-0287

If disposing of this product, please recycle the paper responsibly.

IntechOpen Book Series
Civil Engineering
Volume 20

Aims and Scope of the Series

Civil engineering is a traditional field of engineering from which most other branches of engineering have evolved. It comprises traditional sub-areas like transportation, structures, construction, geotechnics, water resources, and building materials. It also encompasses sustainability, risk, environment, and other concepts at its core. Historically, developments in civil engineering included traditional aspects of architecture and urban planning as well as practical applications from the construction industry. Most recently, many elements evolved from other fields of knowledge and topics like simulation, optimization, and decision science have been researched and applied to increase and evolve concepts and applications in this field. Civil engineering has evolved in the last years due to the demands of society in terms of the quality of its products, modern applications, official requirements, and cost and schedule restrictions. This series addresses real-life problems and applications of civil engineering and presents recent, cutting-edge research as well as traditional knowledge along with real-world examples of developments in the field.

Meet the Series Editor

Professor Assed N. Haddad is a Civil Engineer with a degree from the Federal University of Rio de Janeiro (UFRJ) earned in 1986, as well as a Juris Doctor degree from the Fluminense University Center earned in 1993, and a Master's degree in Civil Engineering from the Fluminense Federal University (UFF) obtained in 1992. He completed his Ph.D. in Production Engineering from COPPE / Federal University of Rio de Janeiro in 1996. Professor Haddad's academic pursuits have taken him to postdoctoral stays at the University of Florida, USA in 2006; at the Universitat Politècnica de Catalunya, Spain in 2010; and at the University of New South Wales Sydney, Australia in 2019. Currently, he serves as a Full Professor at the Federal University of Rio de Janeiro. He has held visiting professorships at various institutions including the University of Florida, Universitat Politècnica de Catalunya, Universitat Rovira i Virgili, and Western Sydney University. His research expertise encompasses Civil, Environmental, and Production Engineering, with a primary focus on the following topics: Construction Engineering and Management, Risk Management, and Life Cycle Assessment. He has been the recipient of research grants from the State of Rio de Janeiro, Brazil: CNE FAPERJ from 2019 to 2022 and from 2023 to 2025. Additionally, his research grants obtained from the Brazilian Government CNP since 2012 last to this date. Professor Haddad has been involved in several academic endeavors, being the Guest Editor of the International Journal of Construction Management; MDPI's Sustainability, Energies, and Infrastructures; Associate Editor at Frontiers in Built Environment / Sustainable Design and Construction; Guest Editor at Frontiers in Built Environment / Construction Management; and Academic Editor of the Journal of Engineering, Civil Engineering Section of Hindawi. He is currently a Professor of the Environmental Engineering Program at UFRJ and the Civil Engineering Program at UFF.

Meet the Volume Editors

Dr. Masoud Mohebbi is an accomplished Railway Engineer and Research Scientist specializing in high-speed train aerodynamics and Computational Fluid Dynamics (CFD). He holds a Ph.D. in Railway Engineering from the Iran University of Science and Technology. Dr. Mohebbi has authored over 30 peer-reviewed publications and has led significant national research projects focused on wind effects on railway structures and vehicle stability. His expertise lies in leveraging advanced computational methods, such as the Lattice Boltzmann Method, to enhance the safety and resilience of high-speed rail systems. He serves as an editor and reviewer for numerous international journals, bridging the gap between theoretical research and practical engineering applications in modern transport infrastructure.

Dr. Rasul Mohebbi is an Associate Professor at Damghan University in Iran and a dedicated researcher in heat transfer and fluid dynamics. His expertise involves advanced Lattice Boltzmann Method (LBM) simulations to study hybrid nanofluids and thermal management strategies. He has conducted significant research on fluid dynamics in trains, focusing on optimizing energy efficiency and cooling systems to enhance performance and sustainability in rail transport. His work reflects a commitment to innovation in thermal sciences, addressing key challenges in both industrial and transportation applications.

Contents

Preface

The global transportation landscape is undergoing a profound transformation, with High-Speed Rail (HSR) at the forefront of the transition toward sustainable, low-carbon mobility. As train speeds increase and networks expand across diverse geographical regions, the engineering challenges have evolved from simple track construction to the management of highly complex, integrated systems. This book, *Innovations in High-Speed Railway Engineering - Aerodynamics, Infrastructure Resilience, and Sustainable Mobility,* provides a comprehensive exploration of the cutting-edge technologies and methodologies shaping the future of this sector.

The volume is organized into three core thematic areas. The first section focuses on Aerodynamics and AI-Driven Safety. The introductory chapter, "Introductory Chapter: The Convergence of Aerodynamics, Digital Resilience, and Sustainable Logistics in High-Speed Rail", sets the stage by synthesizing the multi-disciplinary nature of modern rail systems. This is followed by "Crosswind Stability of High-Speed Train in Unsteady Wind Conditions," which explores integrating machine learning into stability assessment. "Crosswind Stability of High-Speed Train in Unsteady Wind Conditions" provides deep insights into the physics of transient wind loads.

The second section examines Infrastructure Resilience, specifically focusing on the railway's "digital nervous system". The chapter "Integrated Track Circuit Technology Based on All-Electronic Interlocking Systems in China", highlights advancements essential for maintaining safety and high capacity in increasingly dense networks. Finally, the third section explores Sustainable Operations and Logistics. The chapter "A Review of Sustainable Location-Allocation Models for City Logistics Using High-Speed Rail: Case Study of the Ottawa-Quebec Corridor and Greater Montreal Area" explores the strategic integration of HSR into urban supply chains.

As the Academic Editors, we would like to express our gratitude to the international team of contributors. We hope this volume serves as a vital resource and catalyst for further innovation in railway engineering.

Masoud Mohebbi
School of Railway Engineering,
Iran University of Science and Technology,
Tehran, Iran

Rasul Mohebbi
School of Engineering,
Damghan University,
Damghan, Iran

Section 1

Aerodynamics, Stability, and AI-Driven Safety

Chapter 1

Introductory Chapter: The Convergence of Aerodynamics, Digital Resilience, and Sustainable Logistics in High-Speed Rail

Masoud Mohebbi and Rasul Mohebbi

1. The evolving paradigm of high-speed rail

The global renaissance of high-speed rail/railway (HSR) stands as a testament to the engineering community's commitment to decarbonizing transportation while fostering economic connectivity. However, as operational speeds push beyond the 300 km/h frontier and network densities increase, the fundamental challenges of railway engineering have shifted. The contemporary objective is no longer merely the pursuit of maximum velocity; rather, it is the holistic integration of complex subsystems to ensure resilience against environmental unpredictability and the optimization of operational efficiency.

This transition is driven by a necessity to manage the intricate interplay between the vehicle and its environment. As trains traverse diverse topographies – from open plains to complex tunnel-embankment sections – the aerodynamic interactions become exponentially more volatile. Simultaneously, the infrastructure governing these movements is undergoing a digital transformation, moving away from electromechanical legacies toward intelligent, data-driven control systems. Furthermore, the economic sustainability of these massive infrastructure investments now hinges on maximizing asset utilization, not only through precision maintenance but also by expanding the utility of HSR networks into the realm of urban logistics.

2. The intersection of aerodynamics and artificial intelligence

The interaction between a high-speed train and the atmosphere remains the primary constraint on safety and stability. Crosswinds, in particular, pose a critical threat, inducing lateral forces and overturning moments that can compromise the wheel–rail contact mechanics. Traditional methodologies, such as wind tunnel experiments and Computational Fluid Dynamics (CFD), have long served as the bedrock of aerodynamic analysis. However, these physics-based approaches are often computationally intensive and struggle to provide the real-time responsiveness required to mitigate transient, stochastic phenomena like sudden gusts or the "pulse impact" of trains passing in tunnels.

IntechOpen

A paradigm shift is currently underway with the integration of artificial intelligence (AI) into railway aerodynamics. The emergence of machine learning (ML) and deep learning (DL) offers a data-driven alternative capable of modeling complex nonlinearities with unprecedented speed. Advanced techniques, such as surrogate modeling and physics-informed machine learning (PIML), allow engineers to predict aerodynamic loads by embedding physical laws into neural network architectures, ensuring that predictions are both fast and physically plausible. Moreover, the concept of the digital twin – a virtual replica of the physical asset continuously updated with sensor data – is revolutionizing safety protocols. By enabling active control systems to adjust suspension and speed in response to real-time gust detection, the industry is moving from a reactive safety stance to a proactive, predictive one [1–3].

3. Digitalization of signaling and infrastructure control

Parallel to the advancements in vehicle dynamics, the nervous system of the railway – its signaling and control infrastructure – is being redefined. The reliability of track circuits, which serve as the fundamental sensors for detecting section occupancy, is paramount for high-density operations. Historically, these systems relied on phase-sensitive circuits and electromechanical relays. While robust, these analog architectures are susceptible to mechanical wear, possess low return coefficients, and are increasingly vulnerable to harmonic interference from modern high-power traction systems.

The industry is now transitioning toward all-electronic interlocking systems and integrated track circuit technologies, such as the ZPW-2000 series. These modern architectures utilize jointless track circuits and microelectronic receivers to create a seamless, unified platform that integrates station and section controls. By employing redundant electronic control modules (often in two-out-of-two configurations) and advanced digital signal processing, modern signaling systems can effectively filter out complex electromagnetic interference. This technological leap not only enhances safety integrity levels (SILs) but also facilitates the high-throughput, bi-directional communication required for future automatic block signaling and communication-based train control (CBTC) [3, 4].

4. Sustainable mobility: Maintenance efficiency and city logistics

The sustainability of HSR systems extends beyond energy efficiency to encompass the longevity of assets and the diversification of their utility. In the domain of maintenance, the adoption of industrial performance metrics, such as overall equipment effectiveness (OEE), is transforming how railway equipment is utilized. Originally developed for manufacturing, OEE quantifies performance through availability, productivity, and quality. When applied to heavy railway maintenance machinery – such as tampers and ballast regulators – this metric reveals hidden inefficiencies in production capacity. By shifting from schedule-based to data-driven maintenance planning, operators can ensure that critical assets are available when needed, thereby enhancing the resilience of the infrastructure.

Furthermore, the role of HSR is expanding to address the critical challenge of urban freight distribution. As urbanization accelerates, city logistics face severe congestion and environmental pressure. The strategic integration of HSR networks for rapid, low-carbon freight transport – known as location-allocation for city logistics – is emerging as a vital solution. Optimizing the location of logistics hubs and the allocation of freight capacity on high-speed lines requires sophisticated mathematical modeling and metaheuristic algorithms. This dual-use approach, combining passenger and freight utility, represents a crucial step toward financially and environmentally sustainable mobility solutions for the future [5–7].

5. Conclusion

The future of HSR engineering lies in the successful convergence of these diverse disciplines. The stability of the vehicle cannot be separated from the intelligence of the signaling system or the efficiency of its maintenance and logistical operations. As the sector moves forward, the synthesis of AI-driven aerodynamic safety, fully electronic infrastructure control, and rigorous, data-centric operational management will define the next generation of HSR. This volume aims to explore these frontiers, offering insights into the technologies that will make tomorrow's railways faster, safer, and more resilient.

Acknowledgment

The authors acknowledge the use of ChatGPT (OpenAI) for language polishing of the chapter.

Author details

Masoud Mohebbi[1*] and Rasul Mohebbi[2]

1 Iran University of Science and Technology, Tehran, Iran

2 School of Engineering, Damghan University, Damghan, Iran

*Address all correspondence to: mohebbi@alumni.iust.ac.ir

References

[1] Baker CJ. A framework for the consideration of the effects of crosswinds on trains. Journal of Wind Engineering and Industrial Aerodynamics. 2013;**123**:130–142. DOI: 10.1016/j.jweia.2013.09.015

[2] Ouyang D-H, Deng E, Yang W-C, Ni Y-Q, Chen Z-W, Zhu Z-H, Zhou G-Y. Nonlinear aerodynamic loads and dynamic responses of high-speed trains passing each other in the tunnel–embankment section under crosswind. Nonlinear Dynamics. 2023;**111**(13):11989–12015. DOI: 10.1007/s11071-023-08479-7

[3] Phusakulkajorn W, Núñez A, Wang H. Artificial intelligence in railway infrastructure: Current research, challenges, and future opportunities. Intelligent Transportation Infrastructure. 2023;**2**. DOI: 10.1093/iti/liad016

[4] Feng H, Xing K, Sun J, Song M. A domestic electric control module for 25 Hz track circuit. Foreign Electronic Measurement Technology. 2021;**10**:60–65

[5] Muchiri P, Pintelon L. Performance measurement using Overall Equipment Effectiveness (OEE): Literature review and practical application. International Journal of Production Research. 2019;**57**(15–16):5026–5043. DOI: 10.1080/00207540601142645

[6] Tan Y, Wang M. Wind tunnel measurement of aerodynamic characteristics of trains passing each other on a simply supported box girder bridge. Railway Engineering Science. 2021;**29**:152–162. DOI: 10.1007/s40534-021-00231-4

[7] Wang Y, Song R, He S, Song Z. Train routing and track allocation optimization model of multi-station high-speed railway hub. Sustainability. 2022;**14**(12):7292. DOI: 10.3390/su14127292

Chapter 2

Crosswind Stability of High-Speed Train in Unsteady Wind Conditions

Dongqin Zhang

Abstract

The safety of high-speed train (HST) operations under unsteady crosswind conditions has become an increasingly critical concern, particularly in complex wind environments such as tunnel exits and tornadic regions. This study proposes a novel evaluation framework to assess the crosswind stability of HSTs in unsteady wind fields. The framework combines aerodynamic force modeling and multibody simulations, and its accuracy is validated through wind tunnel experiments. To better quantify the transient aerodynamic responses, a dynamic amplification factor (DAF) and a characteristic wind curve (CWC) are introduced for HSTs. These parameters effectively capture the intensity and duration of unsteady wind loads and their amplification on vehicle dynamics. Furthermore, similar DAF and CWC formulations are developed for HSTs subjected to tornadic winds, highlighting the extreme vulnerability of trains to rapidly varying high-intensity wind forces. Based on these findings, three countermeasures are proposed to enhance operational safety: (1) an early warning system for turbulent winds, (2) a windbreak design with spatially varying porosity to mitigate train responses in tunnel exit winds, and (3) a tornado detection and warning system specifically tailored for HSTs. The comprehensive methodology and mitigation strategies presented in this study provide a robust scientific basis for improving the crosswind resilience of HSTs. This research offers essential insights for infrastructure design and operational planning, contributing to safer and more reliable high-speed rail systems in wind-prone regions.

Keywords: crosswind stability, aerodynamics, dynamics, high-speed train, tunnel exit wind, tornadic wind

1. Introduction

With the continued expansion of high-speed rail networks worldwide, ensuring the safety and stability of high-speed trains (HSTs) under complex environmental conditions has become a critical engineering challenge. Among the various external disturbances, crosswinds pose a significant threat to the operational safety of HSTs, particularly due to their potential to induce large aerodynamic forces and moments that may lead to derailment or instability. In fact, recent derailment accidents in China, Japan, Belgium, and Switzerland have been attributed to crosswind-induced overturning [1]. While many previous studies have focused on steady wind conditions, real-world wind environments are inherently unsteady – for example, sudden gusts, tunnel exit winds (TEWs), and tornadic winds. Understanding the impact of unsteady crosswinds on train dynamics is essential for developing reliable safety assessment methods and mitigation strategies.

One common form of unsteady wind is gust wind, which can arise from atmospheric turbulence or wind interaction with topography and infrastructure. Gusts are characterized by rapid fluctuations in wind speed and direction over short durations. For instance, the European railway standard EN 14067–6 defines a "Chinese hat" wind gust to represent an idealized gust pulse [2]. For HSTs, these transient wind loads can result in sudden and significant variations in aerodynamic forces, potentially triggering unstable vehicle responses, particularly when trains operate on elevated tracks or open terrain. The dynamic characteristics of gusts require a time-dependent analysis framework to capture the interaction between the aerodynamic loads and the vehicle's dynamic response accurately.

When a HST exits a tunnel into open terrain, it experiences a sudden change in airflow known as a TEW, where the sudden change creates a transient wind field. As the train emerges from the tunnel, the aerodynamic forces and vehicle motions increase rapidly – a behavior totally different from what occurs under steady or turbulent winds [3]. In practice, this means the lateral wind load can spike and then oscillate as the train clears the tunnel. Experiments and simulations confirm that this can cause greater wheel lift (unloading) than predicted by steady-wind analysis. The spatial and temporal complexity of tunnel exit winds necessitates the use of unsteady aerodynamic models and refined simulation techniques to fully understand their effects.

In addition to atmospheric gusts and tunnel exit phenomena, extreme weather events such as tornadoes present severe unsteady wind conditions. Tornadic winds are highly non-uniform, rotating, and can exert rapidly varying lateral forces on trains. Investigations have noted that several past railway accidents (e.g., on the Touzai, Uetsu, and Nippo lines in Japan, and an Illinois freight train incident in the United States) are suspected to have involved tornadic winds [4, 5]. Though relatively rare, the catastrophic impact of such events, illustrated by past derailment accidents, demands thorough investigation. Studying the aerodynamic interaction between tornado-like vortices and train structures can provide critical insights into emergency response protocols and structural resilience under extreme loading.

In summary, the crosswind stability of HSTs under unsteady wind conditions is a multifaceted problem that requires comprehensive analysis across a range of wind scenarios. This chapter focuses on three representative unsteady wind types—gust winds, tunnel exit winds, and tornadic winds—and explores their distinct mechanisms and effects on train stability. By investigating these cases, the chapter aims to enhance the understanding of unsteady wind–train interactions and support the development of more robust safety and design guidelines for high-speed rail systems operating in diverse environmental conditions.

2. State of the art

2.1 Field test

Field testing plays a crucial role in advancing the understanding of aerodynamic and dynamic responses of HSTs under realistic environmental conditions. While wind tunnel experiments and numerical simulations offer controlled environments for isolated analyses, they often rely on simplified assumptions that may not fully capture the complexity of real-world scenarios. In contrast, field tests provide direct measurements of train responses under actual operational conditions, including natural wind environments, track irregularities, and train–infrastructure interactions. Through the deployment of sensors on train bodies, bogies, and wheels, key parameters such as lateral acceleration, wheel unloading ratio, and aerodynamic pressure distributions can be captured in real time.

In Japan, a stationary train positioned near the coast of Hokkaido was used to examine wind-induced forces under natural conditions [6]. As shown in **Figure 1**, aerodynamic loads on different cars, along with wind speed and direction, were

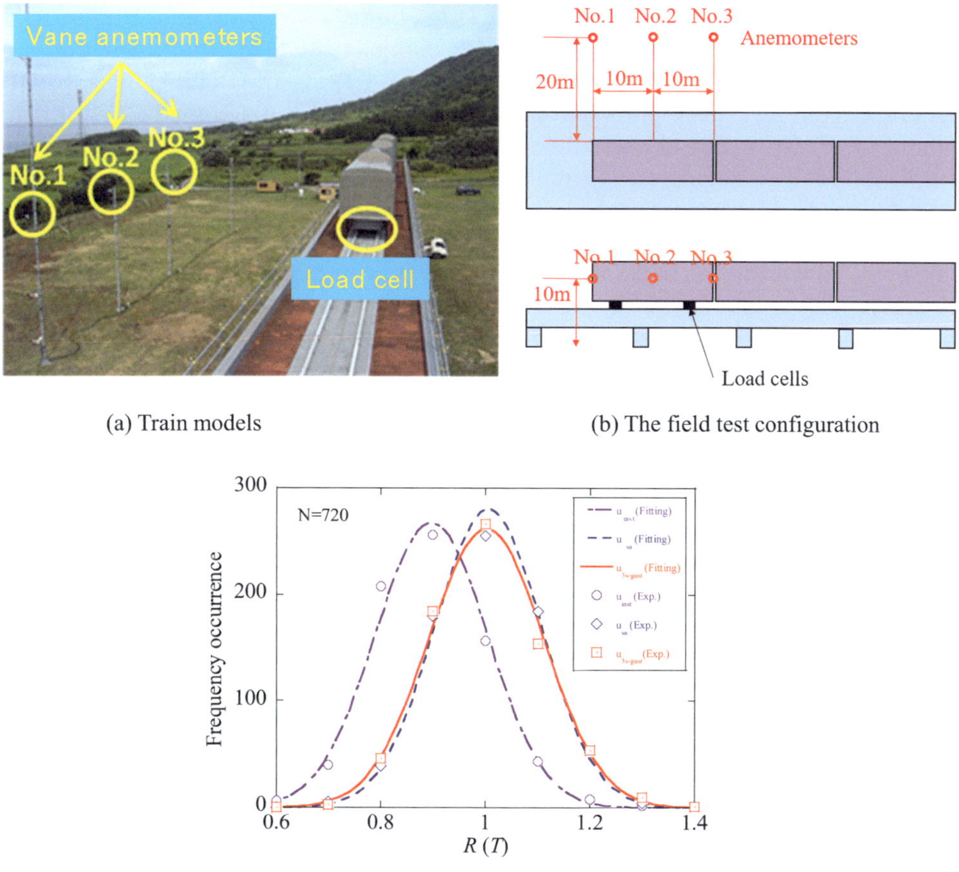

(a) Train models

(b) The field test configuration

(c) Frequency distribution of the ratio of the maximum measured and calculated overturning moment

Figure 1.
The field test for the unsteady aerodynamic forces acting on a railway vehicle in crosswinds.

recorded every 10 seconds. Based on this dataset, Nagumo et al. [7, 8] proposed using 3-second averaged wind speeds to more reliably estimate overturning moments caused by crosswinds. This approach captures the spatial coherence of wind acting along the carriage's length. **Figure 1** presents the distribution of the ratio between measured and estimated overturning moments. Here, u_{inst}, u_{sa}, and $u_{3s-gust}$ represent the instantaneous, spatial-averaged, and 3-second gust speeds, respectively. The results show that calculations based on 3-second gusts align well with spatial-averaged values, indicating that this method accounts for the size effect of crosswind loading on trains.

On the Network Rail New Measurement Train (NREMT), high-frequency pressure data were collected using pitot and static sensors along the nose and body. Sampling at 128 Hz enabled detailed analysis of gust responses. The correlation between yaw angle and side force coefficient varied with trackside topography, being higher in open terrains and lower in urban areas. In clear surroundings, the relationship between the side force and yaw angle closely matched wind tunnel data, validating the train's aerodynamic response under real gust events [9].

Full-scale field tests in China have provided valuable insights into the aerodynamic and dynamic responses of HSTs under crosswind conditions [10]. Wind speeds along the track were recorded using a vehicle-mounted anemometer to serve as a reference for evaluating aerodynamic loads and vehicle behavior. To enhance wind measurement accuracy, Sun et al. [11, 12] developed a cylindrical anemometer featuring 12 pressure sensors that capture directional wind information. The device was validated through wind tunnel tests across a wide velocity range (5–60 m/s) and further verified in field environments for robustness and precision. Aerodynamic loads such as side force, lift, and rolling moment, as well as car body accelerations, were measured under various wind scenarios. These data enabled the assessment of wheel unloading ratios and overturning tendencies. Because direct measurement of wheel loads at high speed is challenging, rolling angle was used as an alternative indicator of train stability. Liu et al. [13] proposed a real-time attitude monitoring method using coordinate transformations, which was validated by comparing calculated roll angles with bogie air spring deflection data, confirming accuracy and repeatability, as illustrated in **Figure 2**. Gao et al. [14] conducted pressure measurements to examine unsteady aerodynamic loads in winds up to 34 m/s. Results showed peak

Figure 2.
The configuration of full-scale measurement of a HST in China [14].

aerodynamic coefficients at yaw angles between 54.5° and 59.6°, and strong correlations between turbulence and load fluctuations. Frequency-domain analysis revealed dominant pressure pulsations below 1 Hz under crosswind conditions. Additional testing indicated pressure amplitude errors under 5% when using pressure tubes less than 8 m in length [15].

2.2 Wind tunnel test

Wind tunnel testing represents a fundamental experimental approach in railway engineering for evaluating the aerodynamic and dynamic characteristics of HSTs. By utilizing scaled models or representative sections within a controlled airstream, key parameters such as drag coefficients, lift and side forces, rolling moments, surface pressure distributions, flow separation patterns, and aeroacoustic signatures can be accurately measured. Furthermore, dynamic responses to gusts, tunnel exit winds, and tornadoes can be experimentally assessed.

Wind tunnel studies have been widely conducted to investigate the aerodynamic behavior of HSTs under crosswind conditions, considering various train types and environmental settings such as ground surfaces, embankments, and bridges [16–21]. These experiments reveal that inflow conditions and track substructures significantly influence aerodynamic coefficients. When evaluating responses under unsteady winds, the concept of aerodynamic admittance becomes critical and should be considered for accurate dynamic force estimation [22]. Moving model tests have also been introduced to evaluate the effect of train speed. Results indicate that the difference between static and moving train models is often minimal under certain conditions [23]. Zou et al. [24] developed a wind tunnel system enabling scaled HST models (1:8–1:30) to travel at speeds of up to 50 m/s for studying vehicle–bridge aerodynamic interaction in **Figure 3**. Wireless pressure sensors were used to capture unsteady surface loads accurately. The findings emphasize that dynamic effects are crucial when crosswinds are present, and static model tests alone may not reflect actual aerodynamic behavior during high-speed operation. Furthermore, crossing scenarios on box girder bridges were examined to evaluate transient aerodynamic forces during train–train encounters. The results show that the sheltering effect significantly alters the loads on the leeward vehicle, depending on wind and train speeds [25].

(a) High-speed train models (b) The launch system

Figure 3.
The launch system for a moving HST model.

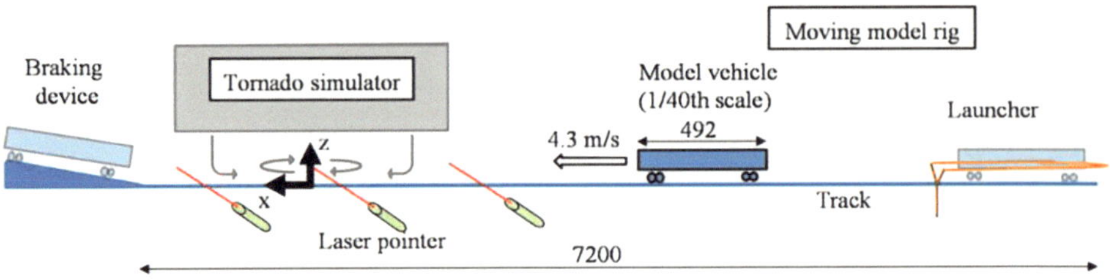

Figure 4.
The wind tunnel test for aerodynamic forces on a railway vehicle in tornadic winds [4].

Recent studies have also focused on tornado-induced loads. Using tornado simulators, both stationary [26] and moving models were tested [4, 27]. Peak lateral forces occur near the tornado core radius, while lift forces peak at the center, with distributions influenced by vortex structure and train length, as shown in **Figure 4** [4].

2.3 Numerical simulation

Numerical simulations based on computational fluid dynamics (CFD) have been extensively applied to analyze the crosswind aerodynamics of HSTs, including the evaluation of drag and lift forces, flow structures, and wake dynamics [28–32]. These studies often focus on complex scenarios, such as a train encountering windbreaks or open-cut trenches, where variations in aerodynamic loads are examined in detail [33]. In particular, the influence of tunnel exit winds has been the subject of multiple investigations, comparing outcomes across various numerical techniques, vehicle geometries, and wind inflow profiles [34, 35]. Xu et al. [36] explored the interaction between a HST and a simulated tornado vortex, capturing the complex aerodynamic loading as the train traverses a rotating wind field. Despite the versatility of CFD tools, challenges persist in accurately resolving unsteady phenomena, such as vortex detachment and reattachment near streamlined train bodies. These difficulties become more pronounced under transient wind conditions, necessitating careful validation against experimental or full-scale test data to ensure the reliability of the predictions.

To complement aerodynamic studies, the dynamic response of railway vehicles to wind-induced loads is often computed using either quasi-static approaches or multibody dynamics (MBD). In Europe, simplified mass-spring models, such as the three- and five-mass representations, are commonly used for practical assessments of vehicle overturning under gusts [37]. MBD methods provide a more detailed representation of vehicle dynamics and have been adopted to assess train stability under synthetic gust models, such as the Chinese hat profile [2, 5, 38].

Japan has developed specific modeling practices, including a quasi-static model with three degrees of freedom and a half-car configuration by Hibino et al. [39], which is widely used to derive characteristic wind curves (CWCs) for commuter trains [40]. A more detailed 31-degree-of-freedom model has also been proposed for simulating the effects of gusty winds [41], although its adoption in the railway industry remains limited. Comparative analyses between simplified models and full multibody simulations have shown reasonable agreement under some conditions, such as Chinese hat gusts [42], suggesting that reduced-order models can offer practical insights when validated properly.

However, discrepancies emerge under highly transient wind environments. For instance, when simulating the impact of tunnel exit winds [3, 43] or tornado-induced gusts [44], quasi-static methods often fail to capture the peak dynamic responses observed in multibody simulations. This underestimation highlights the importance of using time-resolved dynamic models, especially for safety evaluations involving unsteady or extreme wind scenarios. In addition, the influence of vehicle configuration on crosswind sensitivity has been studied, showing that parameters such as car body shape, mass distribution, and suspension design play a critical role in determining overall stability [45].

3. Methods

3.1 Wind scenarios and aerodynamic forces

The spatial distribution of steady wind, tunnel exit wind [3], and tornadic wind [5, 46, 47] is illustrated in **Figure 5**. The variation of tunnel exit wind is described by the one-minus-gust wind model, as shown in previous studies [48]. The Burgers-Rott model is adopted for the wind distribution.

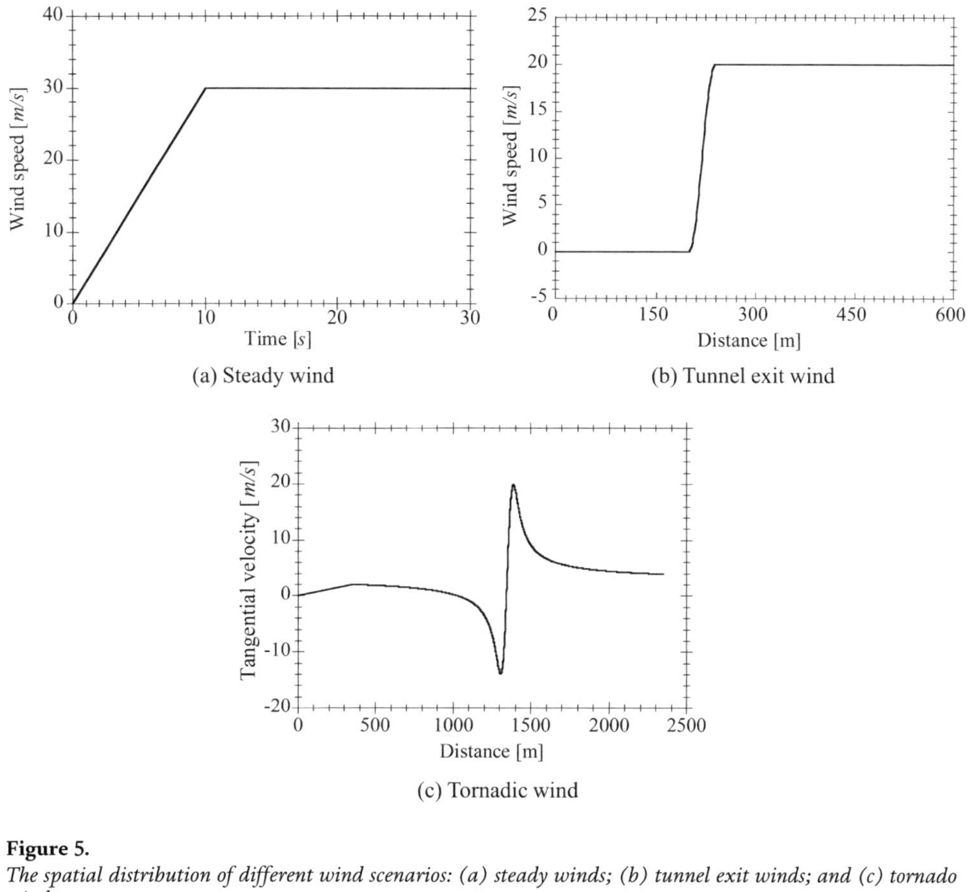

Figure 5.
The spatial distribution of different wind scenarios: (a) steady winds; (b) tunnel exit winds; and (c) tornado winds.

$$U = \frac{1}{2}\Delta U\left(1 - cos\frac{2\pi x}{2L_d}\right) \tag{1}$$

where ΔU and L_d are gust amplitude and gust duration.

$$V = \frac{V_{Max.S}}{0.72}\frac{R_{Max}}{r}\left\{1 - exp\left[-1.26\left(\frac{r}{R_{Max}}\right)^2\right]\right\} \tag{2}$$

$$V_T = cV_{Max.T}, \quad V_{Max.T} = V_{Max.S} + V_T \tag{3}$$

where $V_{Max.S}$ and $V_{Max.T}$ are the maximum tangential velocities of stationary and translational tornadoes, R_{Max} presents the core radius, V_T is the tornado translational velocity, and $c = 0.15$ is obtained from statistical data in Japan.

The spatial wind speed distributed along the track is first transformed into the temporal wind speed, and then aerodynamic forces are calculated using the quasi-steady theory.

$$v_c(t) = \sqrt{\frac{\int_{\left(v_{tr}t - \frac{L_0}{2}\right)}^{\left(v_{tr}t + \frac{L_0}{2}\right)} v_{wi}^2 dx}{L_0}}, \quad i = 1 \cdots N \tag{4}$$

where V_{tr} is the train velocity, L_0 is the length of one carriage, and v_{wi} is the local wind speed at a discrete section.

The resultant wind speed $v_a(t)$ and the angle of attack for the resultant wind speed $\beta(t)$ as shown in **Figure 6(a)** are calculated using the temporal wind speed $v_c(t)$ and are expressed as follows:

$$v_a(t) = \sqrt{[V_{tr} + v_c(t)cos\beta_w]^2 + [v_c(t)sin\beta_w]^2} \tag{5}$$

$$\beta(t) = arctan\left[\frac{v_c(t)\ sin\beta_w}{V_{tr} + v_c(t)\ cos\beta_w}\right] \tag{6}$$

where β_w represents the angle of attack for the wind.

The side force $F_S(t)$, lift force $F_L(t)$, and rolling moment $M_R(t)$as shown in **Figure 9(b)** are calculated using the quasi-steady theory and are presented as follows:

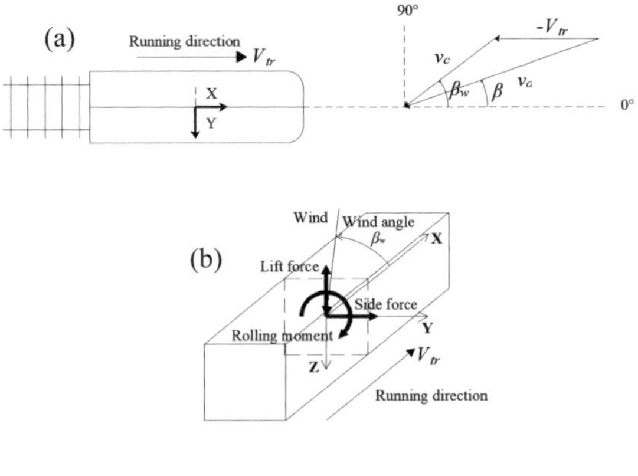

Figure 6.
The directions of (a) crosswind and (b) aerodynamic forces.

$$F_S(t) = \frac{1}{2}\rho A C_s\left(\beta(t)\right)v_a^2(t) \tag{7}$$

$$F_L(t) = \frac{1}{2}\rho A C_L\left(\beta(t)\right)v_a^2(t) \tag{8}$$

$$M_R(t) = \frac{1}{2}\rho A C_M\left(\beta(t)\right)v_a^2(t)H_0 \tag{9}$$

where ρ is air density, A is the side area of the carriage, C_s, C_L, C_M are aerodynamic coefficients for side force, lift force, and rolling moment, H_0 is the height of the car body.

3.2 Multibody simulation

A multibody model of a HST is developed, comprising a car body, two bogies, and four wheelsets, as illustrated in **Figure 7**. The connections between the wheelsets, bogies, and car body are represented by longitudinal, lateral, and vertical springs and dampers, simulating both primary and secondary suspension systems. Additionally, nonlinear components, such as lateral bump stops and anti-snake movement dampers, are incorporated between the bogies and car body to capture realistic dynamic interactions. The contact forces between wheels and rails are modeled using Hertzian nonlinear elastic contact theory.

Each structural component is treated as a rigid body, and the system's topology is defined accordingly. While the equations of motion can be derived using Newton's second law, the complexity of the system necessitates a more efficient formulation. To automate the modeling of such large-scale mechanical systems, the equations are expressed as a set of implicit differential-algebraic equations (DAEs).

These DAEs are numerically solved using the backward differentiation formula method. The time domain is discretized with a uniform time step $h_i = t_{i+1} - t_i$. At each time step t_{n+k}, the derivative of a state variable $x(t)$ is approximated using previously computed values $x_n = x(t_n), \ldots, x_{n+k} = x(t_{n+k})$. This is achieved by fitting

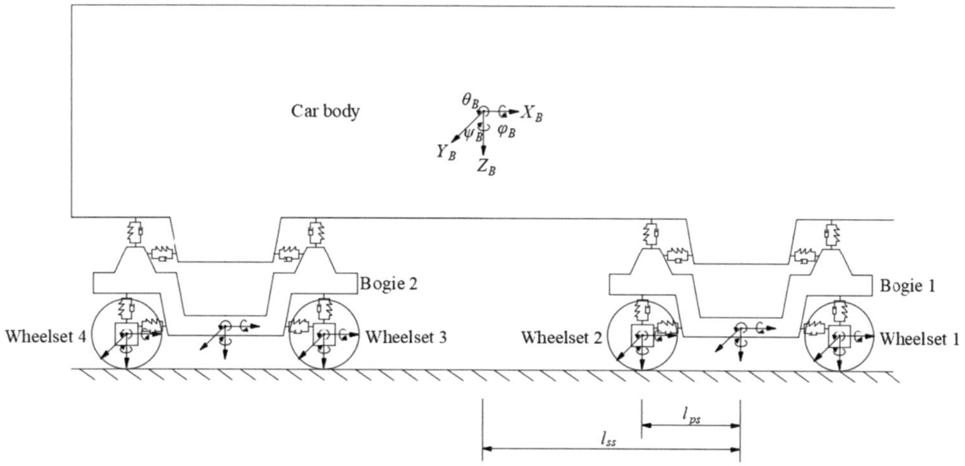

Figure 7.
A HST model.

a polynomial $p \in P_k$ that interpolates the discrete data points, allowing accurate time integration of the system's dynamics.

3.3 Dynamic amplification factor

The dynamic amplification factor (DAF), commonly used in structural engineering, is defined as the ratio between the maximum dynamic response and its static counterpart [49]. To assess the crosswind stability of railway vehicles, metrics such as the overturning coefficient or wheel unloading ratio are employed to quantify the extent of wheel unloading. In this study, the maximum dynamic overturning coefficient is obtained through MBS simulations, while the corresponding static value is derived from steady-state analysis, as illustrated in **Figure 8**. Based on these values, the DAF is expressed as follows:

$$DAF = \frac{max\{D_d\}}{D_s} \tag{10}$$

where D_d, D_s represent the dynamic and static overturning coefficients, respectively.

3.4 Characteristic wind curve

The CWC serves as a critical reference in evaluating the crosswind stability of HSTs. It defines the relationship between the running speed of the train and the allowable crosswind velocity under which safety criteria, such as wheel unloading or derailment risks, remain within acceptable limits. Traditionally, the CWC is derived under the assumption of steady-state wind conditions and quasi-static aerodynamic responses. However, in real-world scenarios, trains are frequently exposed to unsteady wind fields, such as gusts, tunnel exit winds, or tornadoes, which can dynamically amplify train responses.

To address this limitation, we propose a novel method for deriving the CWC that explicitly accounts for unsteady wind effects by incorporating the maximum DAF. The DAF captures the ratio between the peak dynamic response and the corresponding quasi-static response of the train under unsteady aerodynamic loads. By introducing this factor, the proposed approach enables a more realistic and conservative estimation of critical wind speeds, particularly under extreme or

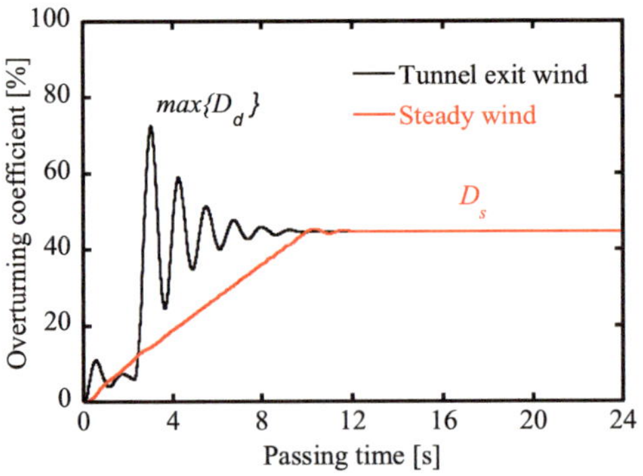

Figure 8.
The definition of the DAF of a HST in unsteady winds.

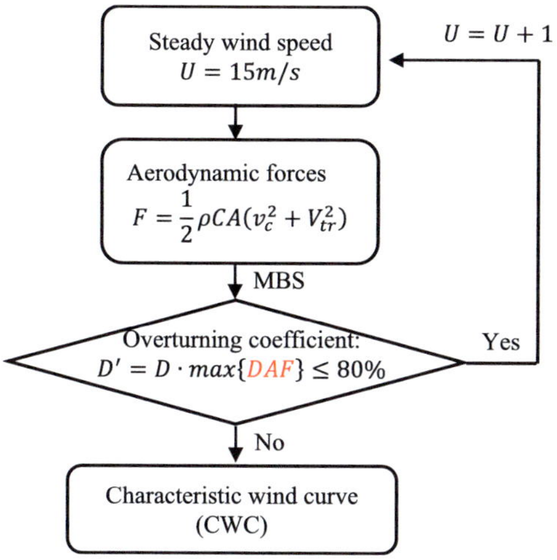

Figure 9.
The process of calculating the CWC of a high-speed train in unsteady winds.

transient wind events, as illustrated in **Figure 9**. This method is implemented through time-domain simulations of vehicle dynamics subjected to synthetic unsteady wind profiles, providing a more robust framework for safety assessment and design guidance in crosswind-prone regions.

4. Results and discussion

4.1 A framework to evaluate the crosswind stability of a HST in unsteady crosswinds

An elastic scale model of a railway vehicle, developed by Hibino et al. [50], is utilized to study the aerodynamic force modeling and dynamic behavior of trains exposed to tunnel exit winds. This 1/10-scale model is based on the 103 series operated by the former Japanese National Railways, as depicted in **Figure 10(a)**. Designed to meet geometric and mechanical similarity laws, the model replicates the prototype's dynamics, with suspension stiffness reduced to 1/100 and component mass scaled down to 1/1000. Running tests were conducted using this model.

The experimental setup consists of three distinct sections: acceleration, testing, and deceleration (**Figure 10(b)**, front car shown). During the acceleration phase, the model vehicle is propelled forward in still air, replicating the motion inside a tunnel. Upon entering the test section, it encounters crosswinds generated by a fan-driven collector, which gradually exposes the side of the front carriage to a growing airflow until it is fully immersed in a uniform wind field, mimicking the sudden exposure to crosswinds at tunnel exits. The train velocity in this region is $V_{tr} = 9.8km/h$, and the crosswind speed is $U_0 = 8.8m/s$ also unspecified. The vehicle then enters the deceleration zone, where braking simulates re-entry into another tunnel. The collector's 3-meter length exceeds the carriage's 2-meter length, ensuring full crosswind exposure before the model transitions out of the test zone.

Figures 10(c)–(e) present the wind field, aerodynamic forces, and roll angle derived from both experiments and the proposed methodology. In the analysis, the

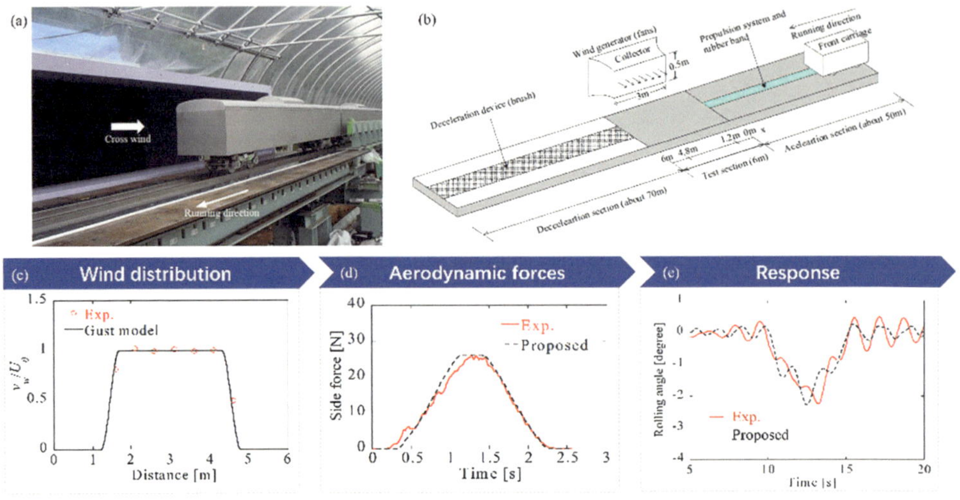

Figure 10.
Validation of the proposed framework using an elastic train model test in a wind tunnel.

spatial wind field is filtered using the equivalent side force method, and aerodynamic loads are computed via quasi-steady assumptions. Multibody simulations are then used to evaluate the train's dynamic response. The strong agreement between experimental and numerical results confirms the framework's accuracy in capturing unsteady crosswind effects.

Based on this validated approach, the DAF of the train can be quantified under various wind intensities and train configurations. The CWC is subsequently determined from the maximum DAF values.

4.2 Crosswind stability of HSTs in tunnel exit winds

The dynamic response of a HST under both tunnel exit winds and steady crosswinds is evaluated, focusing on the influence of gust amplitude, gust duration, and train speed on the wheel unloading ratio. As shown in **Figure 11**, the wheel unloading ratio generally increases with wind speed, followed by noticeable oscillations after reaching peak wind velocity. **Figure 11(a)** indicates that larger gust amplitudes tend to amplify these oscillations, whereas longer gust durations exhibit a stabilizing effect. In particular, when the gust duration is 160 m, the increase in the wheel unloading ratio becomes gradual with minimal fluctuation, and the results under tunnel exit winds and steady winds converge (**Figure 11(b)**). **Figure 11(c)** highlights the role of train speed, revealing that higher speeds intensify the dynamic response. This trend aligns with findings by Ishihara et al. [3] in their investigation of commuter rail systems under crosswind conditions.

To systematically assess the train's sensitivity to various parameters, a series of simulations covering different gust durations, amplitudes, and speeds were conducted, as summarized in **Table 1**. The results are synthesized in **Figure 12**, where the ratio of wheel unloading under tunnel exit winds to that under steady winds is used to derive the DAF. As shown in **Figure 12(a)**, the DAF increases nearly linearly as the wind's passing time decreases. Notably, the DAF becomes significant

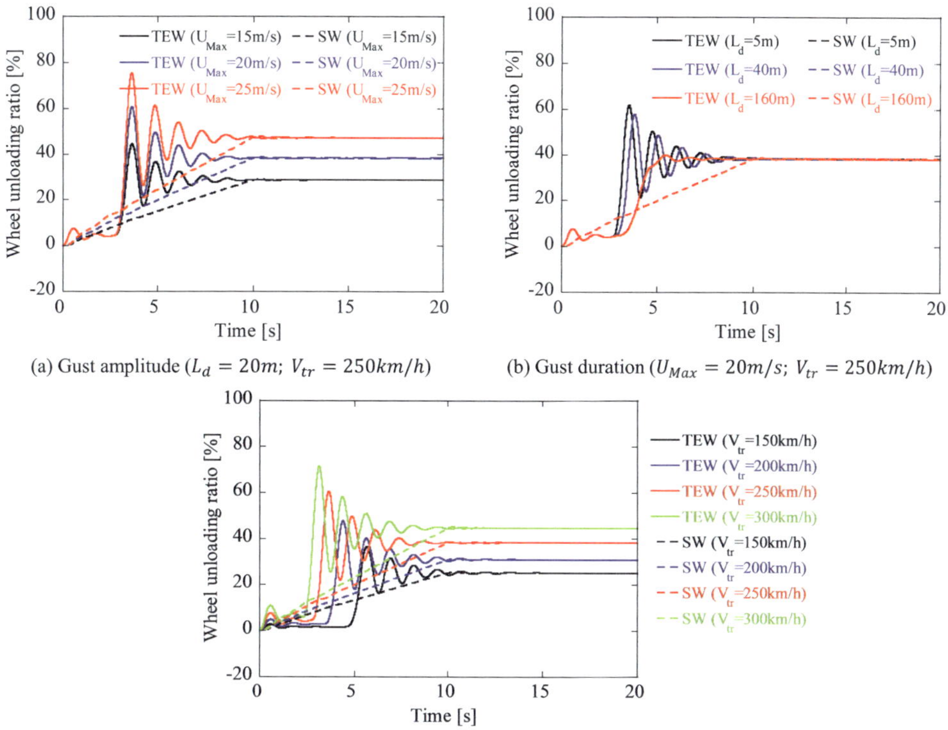

(a) Gust amplitude ($L_d = 20m;\ V_{tr} = 250km/h$) (b) Gust duration ($U_{Max} = 20m/s;\ V_{tr} = 250km/h$)

(c) Train speed ($U_{Max} = 20m/s;\ L_d = 20m$)

Figure 11.
Main parameter effects on wheel unloading ratio of the HST in tunnel exit winds.

Case	L_d [m]	U_{Max} [m/s]	V_{tr} [km/h]
1	5	15	150 ~300 at intervals of 25
2	5	20	150 ~300 at intervals of 25
3	5	25	150 ~300 at intervals of 25
4	20	15	150 ~300 at intervals of 25
5	20	20	150 ~300 at intervals of 25
6	20	25	150 ~300 at the interval of 25
7	40	15	150 ~300 at intervals of 25
8	40	20	150 ~300 at intervals of 25
9	40	25	150 ~300 at intervals of 25
10	80	15	150 ~300 at intervals of 25
11	80	20	150 ~300 at intervals of 25
12	80	25	150 ~300 at the interval of 25
13	160	15	150 ~300 at intervals of 25
14	160	20	150 ~300 at intervals of 25
15	160	25	150 ~300 at intervals of 25

Table 1.
Tunnel exit wind parameters.

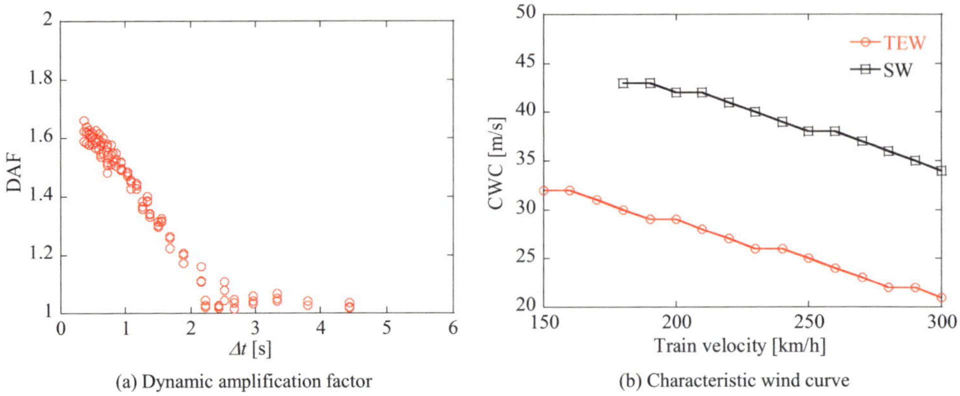

Figure 12.
The DAF and CWC of the HST in tunnel exit winds.

at around 2.5 seconds for the HST. The maximum DAF observed is approximately 1.66.

Figure 12(b) presents the CWC for the HST. It is evident that steady wind conditions correspond to higher characteristic wind speeds, implying reduced dynamic response. In contrast, tunnel exit winds lower the characteristic wind speed by about 13 m/s, indicating greater vulnerability to aerodynamic excitation in these unsteady conditions, which is consistent with earlier studies [3].

4.3 Crosswind stability of HSTs in tornadic winds

The effects of several critical parameters, namely maximum tangential velocity, core radius, and train speed, on the overturning behavior of HSTs in tornado winds are illustrated in **Figure 13**. Solid lines represent results under tornado conditions, while dashed lines correspond to steady wind cases, with the steady wind speed matched to the peak tangential velocity experienced at the train's center. Results clearly show that the overturning coefficient under tornado winds exceeds that under steady winds, primarily due to the sharp gusts characteristic of tornado events. As tangential velocity declines, the overturning effect gradually weakens, approaching zero. Among the parameters, maximum tangential velocity has the most significant impact. As it increases, the difference between tornado and steady wind responses widens.

The core radius also plays a decisive role. With larger core radii, changes in tangential velocity become more gradual, reducing the dynamic amplification effects. Consequently, the overturning coefficients for tornado and steady winds begin to converge. Similarly, train speed strongly influences the response. Lower train velocities are associated with reduced overturning coefficients and a smaller discrepancy between tornado and steady wind conditions. This is partly due to the longer duration of wind exposure at lower speeds, which diminishes the transient impact of tornado gusts.

To quantify these transient effects, the DAF is computed using tornado parameters, including core radius and peak tangential velocity, as summarized in **Table 2**. **Figure 14(a)** presents the DAF results derived from the tornadic winds. The Burgers-Rott model yields a smoothly increasing DAF that peaks at 1.9 as the wind passage time approaches 2.5 seconds.

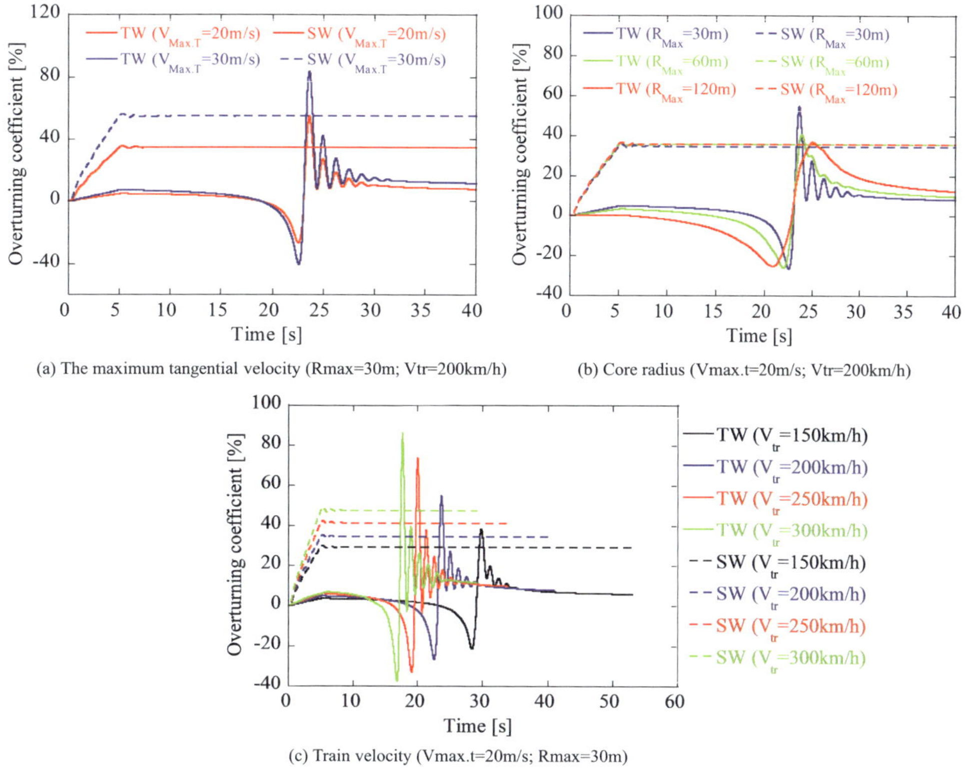

Figure 13.
Main parameter effects on the wheel unloading ratio of the HST in tornadoes.

Case	R_{Max} [m]	$V_{Max.T}$ [m/s]	V_{tr} [km/h]
1	30	20	150-250 at interval of 25
2	30	30	150-250 at interval of 25
3	40	20	150-250 at interval of 25
4	40	30	150-250 at interval of 25
5	50	20	150-250 at interval of 25
6	50	30	150-250 at interval of 25
7	60	20	150-250 at interval of 25
8	60	30	150-250 at interval of 25
9	120	20	150-250 at interval of 25
10	120	30	150-250 at interval of 25

Table 2.
Tornado parameters.

Figure 14(b) shows the CWC for HSTs under tornado winds, along with a comparison to steady wind conditions. The results indicate a significantly lower CWC under tornado scenarios, implying an increased risk. Even for weak tornadoes (e.g., F0 with peak velocities of 17–33 m/s), the characteristic wind

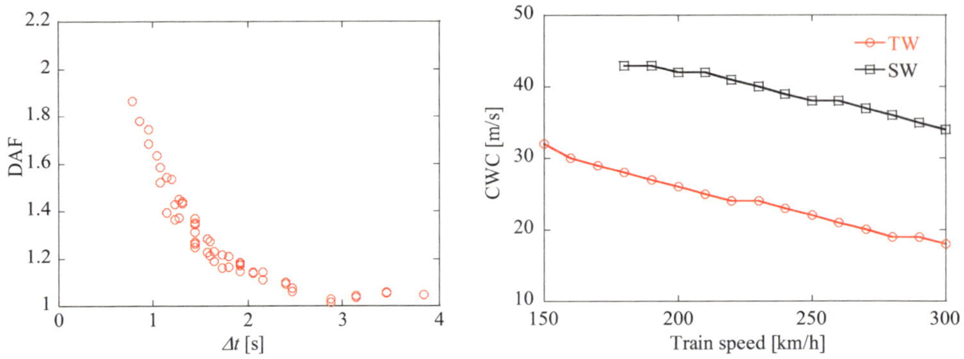

(a) Dynamic amplification factor

(b) Characteristic wind curve

Figure 14.
The DAF and CWC of the HST in tornadic winds.

speed drops below 30 m/s, posing a threat to trains traveling above 150 km/h. Trendlines fitted to the results reveal a nearly linear relationship between CWC and train speed.

5. Countermeasures and design recommendations

5.1 Early warning system

To ensure the safe operation of high-speed rail systems under hazardous wind conditions, it is essential to combine advanced structural and aerodynamic designs with real-time monitoring and decision-making tools. In response to this need, an early warning system has been developed to provide timely alerts and operational guidance when crosswind conditions pose a safety risk. This system integrates wind speed forecasting with the CWC of HSTs along various railway segments. When predicted or observed wind speeds exceed defined safety thresholds, the system triggers appropriate measures such as speed restrictions, temporary service suspensions, or route changes.

Accurate wind speed forecasting is central to the effectiveness of this warning system. To enhance prediction performance, a hybrid approach using variational mode decomposition (VMD) and graph neural networks (GNN) is adopted. VMD enhances the representation of nonstationary meteorological signals, while GNN captures spatial dependencies across meteorological variables. Building on this, a novel deep learning framework is proposed that models pairwise relationships among atmospheric variables to further improve forecasting accuracy, as shown in **Figure 15**.

In addition to temporal predictions, spatial wind speed forecasting along the railway route is also crucial. To achieve this, a spatio-temporal forecasting strategy is introduced, combining deep learning and mesoscale meteorological models. As illustrated in **Figure 16**, observed wind data are first processed by the hybrid deep learning model to predict wind speed and direction over time. These outputs are then used to drive a microscale meteorological model, enabling wind speed estimation at any location within the target region.

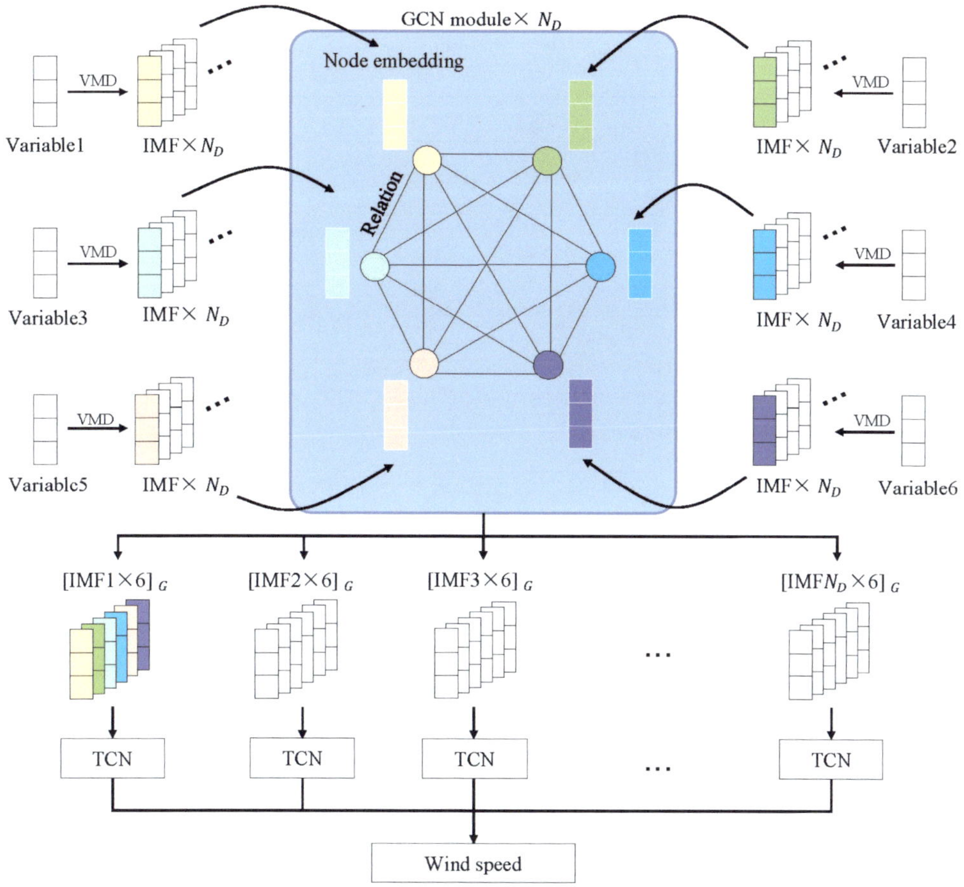

Figure 15.
The structures of a novel hybrid deep learning model considering multiple atmospheric variables.

5.2 Windbreak wall with varying porosities

This study aims to design an effective windbreak configuration to mitigate the maximum unsteady crosswind response of HSTs in areas of intensified wind near tunnel exits. A simple yet adaptable windbreak wall concept is adopted, allowing for future exploration of more complex configurations. The proposed design incorporates five segments with varying porosities ranging from 0% to 80%, increasing in 20% increments toward the downstream end, as illustrated in **Figure 17**. Each segment length (d1–d5) is treated as a design variable, enabling flexible configuration of porosity distributions.

To address the pronounced aerodynamic response often observed at the downstream end of windbreaks, where gust velocities rise sharply, the effectiveness of different porosity layouts is systematically evaluated. **Table 3** outlines 17 windbreak scenarios. Case 1 represents the baseline with a fully solid wall (0% porosity). Cases 2 to 5 employ four levels of porosity (0% to 80%) distributed evenly. Gradient porosity configurations are tested in Cases 6–9 (0%–60%), 10–13 (0%–40%), and 14–17 (0%–20%). Each configuration is assessed under three train speeds: 150, 200, and 250 km/h.

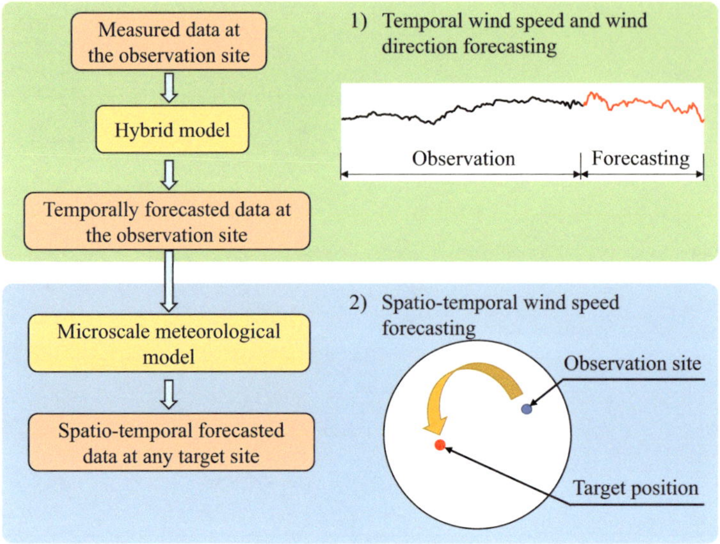

Figure 16.
The structures of a novel spatio-temporal wind speed forecasting.

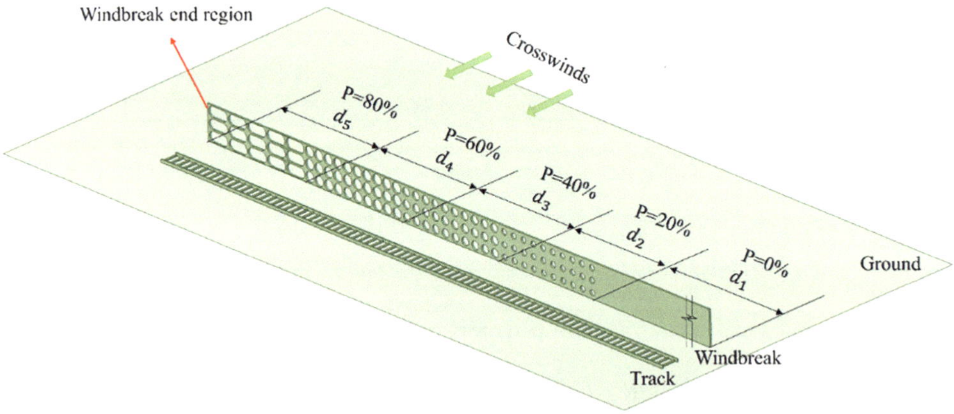

Figure 17.
A newly designed windbreak wall with variable porosities.

Figure 18 summarizes the resulting overturning coefficients across all scenarios. The data reveal a clear reduction in the maximum unsteady crosswind response when variable-porosity windbreaks are used. The greatest improvement, up to 25%, is achieved with the full porosity gradient (0%–80%) at a train speed of 250 km/h. Reductions of approximately 15%, 10%, and 3% are observed for narrower gradients of 0%–60%, 0%–40%, and 0%–20%, respectively.

These findings confirm the effectiveness of variable-porosity windbreaks in attenuating aerodynamic loads at critical locations along the track. The study recommends their application in high-risk areas with strong crosswinds to enhance the operational safety of HSTs.

	d_1 [m] P = 0%	d_2 [m] P = 20%	d_3 [m] P = 40%	d_4 [m] P = 60%	d_5 [m] P = 80%	**Train speed [km/h]**
1	300	–	–	–	–	150/200/250
2	300	10	10	10	10	150/200/250
3	300	20	20	20	20	150/200/250
4	300	30	30	30	30	150/200/250
5	300	40	40	40	40	150/200/250
6	300	13.3	13.3	13.3	–	150/200/250
7	300	26.7	26.7	26.7	–	150/200/250
8	300	40	40	40	–	150/200/250
9	300	53.3	53.3	53.3	–	150/200/250
10	300	20	20	–	–	150/200/250
11	300	40	40	–	–	150/200/250
12	300	60	60	–	–	150/200/250
13	300	80	80	–	–	150/200/250
14	300	40	–	–	–	150/200/250
15	300	80	–	–	–	150/200/250
16	300	120	–	–	–	150/200/250
17	300	160	–	–	–	150/200/250

Table 3.
Windbreak scenarios with different porosities.

5.3 Doppler radar warning system

In tornado-prone regions, ensuring the safety of railway operations presents a critical challenge due to the sudden onset and destructive power of tornadoes. To mitigate the risk of tornado-induced accidents, a Doppler radar-based early warning system specifically designed for railway applications is proposed. This system leverages real-time weather monitoring and predictive decision-making to enhance operational safety during extreme convective weather events.

The system consists of three key components: detection, tracking, and response. First, tornadoes are detected using high-resolution Doppler radar capable of capturing the rotational velocity and core structure of mesocyclones and tornado vortices. Once detected, the system continuously tracks the tornado's movement, estimating its trajectory, speed, and potential impact area using advanced radar analysis and nowcasting techniques. Finally, based on the predicted path and timing of the tornado relative to the railway network, the system initiates real-time decision-making protocols. These include issuing alerts to train drivers and operators, enforcing speed restrictions, halting train operations, or evacuating vulnerable segments if necessary.

This system builds upon operational experiences from the JR East Railway Company in Japan, where a similar tornado detection and response mechanism has been deployed. By integrating meteorological data with railway network operations,

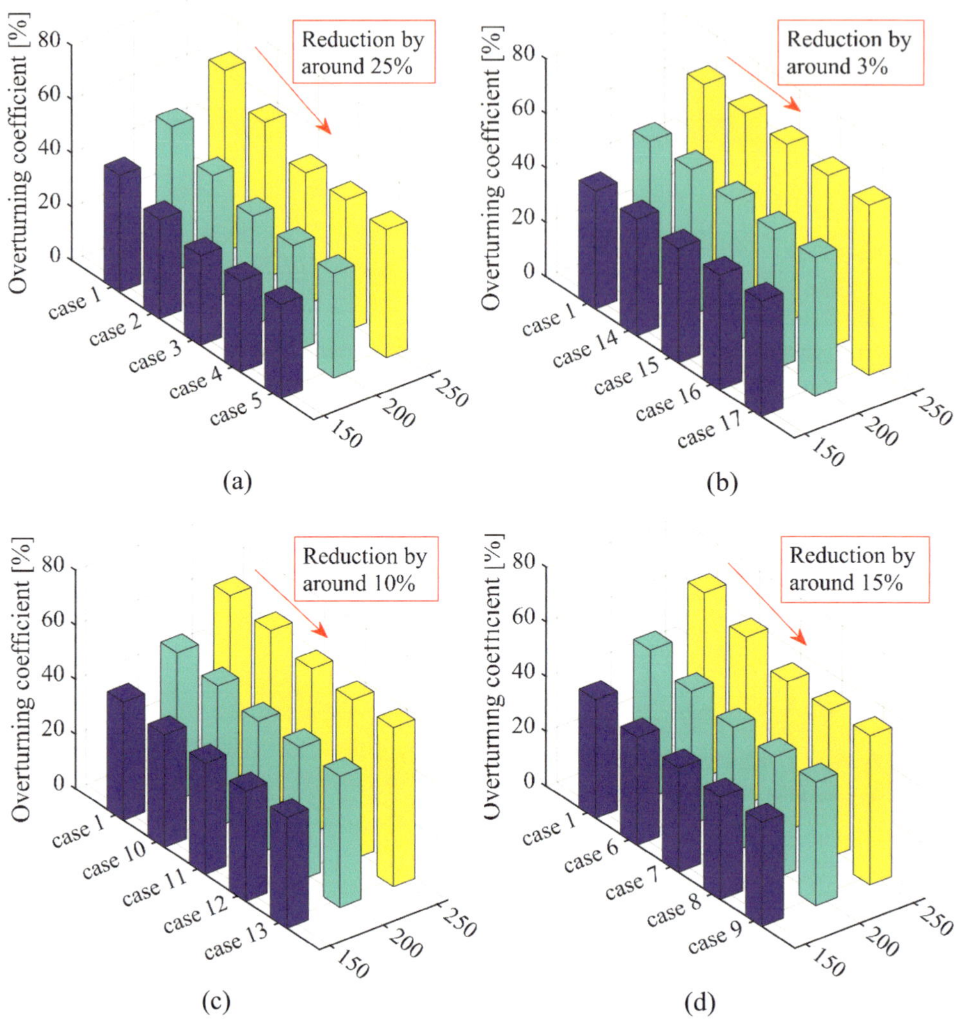

(a) (b) (c) (d)

Figure 18.
Comparison of the maximum overturning coefficient of HST model under different windbreak scenarios.

the system provides a robust defense against severe weather hazards and offers a proactive approach to maintaining safety and service reliability during tornado events.

6. Conclusions and future work

This study proposes a novel and comprehensive framework for evaluating the crosswind stability of HSTs operating under unsteady wind conditions. Recognizing the complexity and variability of such winds, particularly those occurring at tunnel exits and during tornado events, this framework integrates advanced aerodynamic modeling, experimental validation, and tailored countermeasure strategies.

To accurately quantify the aerodynamic forces acting on HSTs in highly transient wind environments, an equivalent side force method was developed and validated through wind tunnel experiments. This method allows for a more realistic representation of unsteady aerodynamic loads, capturing the dynamic nature of winds at tunnel exits and within tornadic vortices. Building on this, two key safety evaluation indicators were introduced: the DAF and the CWC. The DAF quantifies the magnification of train responses due to dynamic wind loading, while the CWC defines the safe operational envelope of the train in terms of allowable wind speeds at different running speeds. By incorporating the maximum DAF into the formulation of the CWC, this framework provides a more conservative and practical tool for assessing the crosswind safety of HSTs in unsteady wind scenarios.

Furthermore, this study identifies and recommends scenario-specific countermeasures to mitigate wind-induced risks. For turbulent or gusty winds, a real-time early warning system is essential to inform operational decisions and reduce accident risk. In tunnel exit zones, windbreak walls with spatially varying porosity are shown to be effective in diffusing strong airflow and reducing sudden side forces on the train. For tornado-prone regions, the implementation of a Doppler radar-based tornado detection and tracking system is critical to enable timely train control interventions and minimize exposure to hazardous vortex-induced loads.

This integrated approach not only improves the understanding of train–wind interactions in unsteady flow conditions but also provides a practical basis for engineering solutions and safety planning. The framework contributes to advancing the resilience and operational safety of high-speed rail systems facing increasingly severe and unpredictable atmospheric disturbances due to climate change.

In the future, firstly, the reliability and generalization of the framework should be further validated under a wider range of operational and environmental conditions. This includes considering different train models, track geometries, and wind profiles observed in various geographic regions. Incorporating long-term field monitoring data and high-fidelity numerical simulations will help to further assess the accuracy and consistency of the equivalent side force method and the proposed DAF and CWC formulations.

Secondly, additional efforts should be directed toward the development and integration of more effective and adaptive countermeasures tailored to different unsteady wind scenarios. For instance, the optimization of windbreak wall designs for tunnel exit winds, considering dynamic porosity and aerodynamic shaping, warrants further investigation. Moreover, the early warning system for gusts and turbulent winds could benefit from real-time data assimilation and AI-based prediction models. For tornado-prone areas, further development of radar-based detection and predictive tracking algorithms, coupled with automated decision-support systems, is essential to minimize operational risks.

In future studies, a multihazard safety management system that integrates detection, forecasting, response, and control strategies is envisioned to promote the safe and resilient operation of high-speed railways under increasingly complex and extreme wind environments.

Conflict of interest

The authors declare no conflict of interests.

Author details

Dongqin Zhang
School of Civil Engineering, Central South University, Changsha, China

*Address all correspondence to: dongqin.zhang@outlook.com

References

[1] Yu M, Jiang R, Zhang Q, Zhang J. Crosswind stability evaluation of high-speed train using different wind models. Chinese Journal of Mechanical Engineering. 2019;**32**:40

[2] EN 14067-6. Railway applications - Aerodynamics - Part6: Requirements and test procedures for cross wind assessment. BS EN. 2010;**14067-6**:2010

[3] Ishihara T, Zhang D, Nagumo Y. Numerical study of dynamic response of railway vehicles under tunnel exit winds using multibody dynamic simulations. Journal of Wind Engineering & Industrial Aerodynamics. 2021;**211**:104556

[4] Suzuki M, Okura N. Study of aerodynamic forces acting on a train using a tornado simulator. Mechanical Engineering Letters. 2016;**2**:1–8

[5] Zhang D, Zhong M, Hu G, Xiao Y. Numerical study of the unsteady crosswind response of high-speed train under local structure-induced unsteady winds by MBS. Engineering Structures. 2023;**281**:115788

[6] Suzuki M, Hibino Y. Field tests and wind tunnel tests on aerodynamic characteristics of train/vehicles under crosswinds. Quarterly Report of RTRI. 2016;**57**:55–60

[7] Nagumo Y, Ishihara T. Reliability of train car against cross winds considering uncertainty in estimated aerodynamic force. Proceedings of the National Symposium on Wind Engineering. 2016;**1**:361–366

[8] Nagumo Y, Ishihara T. A study of crosswind resistant performance of train cars considering uncertainties of acting forces. Journal of Wind Engineering. 2020;**45**:79–92

[9] Gallagher M, Morden J, Baker C, Soper D, Quinn A, Hemida H, et al Trains in crosswinds – Comparison of full-scale on-train measurements, physical model tests and CFD calculations. Journal of Wind Engineering & Industrial Aerodynamics. 2018;**175**:428–444

[10] Liu T, Yan H, Gao H, Xu H, Chen X, Zhang J. Advancements in crosswind safety for multiple-unit train operations on the Southern Xinjiang Railway, China. Advances in Wind Engineering. 2025;**2**:100023

[11] Sun B, Zhou W, Yuan M, Fang E, Gan N. A cylindrical vehicle-mounted anemometer based on 12 pressure sensors—principle, prototype design, and validation. IEEE Sensors Journal. 2018;**18**: 6954–6961

[12] Sun B, Chen G, Chen J, Li X, Tang M, Zhong M. Performance of a vehicle-mounted anemometer under crosswind: Simulation and experiment. Transportation Safety and Environment. 2023;**5**:tdac053

[13] Liu D, Lu Z, Cao T, Li T. A real-time posture monitoring method for rail vehicle bodies based on machine vision. Vehicle System Dynamics. 2017;**55**:853–874

[14] Gao H, Liu T, Gu H, Jiang Z, Huo X, Xia Y, et al. Full-scale tests of unsteady aerodynamic loads and pressure distribution on fast trains in crosswinds. Measurement. 2021;**186**:110152

[15] Liu T, Wang L, Chen Z, Gao H, Li W, Guo Z, et al. Study on the pressure pipe length in train aerodynamic tests and its applications in crosswinds. Journal of Wind Engineering & Industrial Aerodynamics. 2022;**220**: 104880

[16] Cheli F, Corradi R, Rocchi D, Tomasini G, Maestrini E. Wind tunnel tests on train scale models to investigate the effect of infrastructure scenario. Journal of Wind Engineering & Industrial Aerodynamics. 2010;**98**: 353–362

[17] Schober M, Weise M, Orellano A, Deeg P, Wetzel W. Wind tunnel investigation of an ICE 3 endcar on three standard ground scenarios. Journal of Wind Engineering & Industrial Aerodynamics. 2010;**98**:345–352

[18] Cheli F, Giappino S, Rosa L, Tomasini G, Villani M. Experimental study on the aerodynamic forces on railway vehicles in presence of turbulence. Journal of Wind Engineering & Industrial Aerodynamics. 2013;**123**:311–316

[19] Tomasini G, Giappino S, Corradi R. Experimental investigation of the effects of embankment scenario on railway vehicle aerodynamic coefficients. Journal of Wind Engineering & Industrial Aerodynamics. 2014;**131**:59–71

[20] Kikuchi K, Suzuki M. Study of aerodynamic coefficients used to estimate critical wind speed for vehicle overturning. Journal of Wind Engineering & Industrial Aerodynamics. 2015;**147**:1–17

[21] Noguchi Y, Suzuki M, Baker C, Nakade K. Numerical and experimental study on the aerodynamic force coefficients of railway vehicles on an embankment in crosswind. Journal of Wind Engineering & Industrial Aerodynamics. 2019;**184**:90–105

[22] Baker C. A framework for the consideration of the effects of crosswinds on trains. Journal of Wind Engineering & Industrial Aerodynamics. 2013;**123**:130–142

[23] Dorigatti F, Sterling M, Baker CJ, Quinn AD. Crosswind effects on the stability of a model passenger train-A comparison of static and moving experiments. Journal of Wind Engineering & Industrial Aerodynamics. 2015;**138**:36–51

[24] Zou S, He X, Wang H, Tang L, Peng T. Wind tunnel experiment on aerodynamic characteristics of high-speed train-bridge system under crosswind. J Traffic Transp Eng. 2018;**20**:132–139

[25] Li X, Tan Y, Qiu X, Gong Z, Wang M. Wind tunnel measurement of aerodynamic characteristics of trains passing each other on a simply supported box girder bridge. Railway Engineering Science. 2021;**29**:152–162

[26] Cao J, Cao S, Ge Y. Experimental investigations of wind load distributions on a high-speed train under tornado-like vortices. In: Proceedings of the 15th International Conference on Wind Engineering; 2019 Sep 1–6; Beijing, China. p. 437–8

[27] Bourriez F, Soper D, Baker C, Sterling M. Physical model measurement of tornado induced forces on trains. In: Proceedings of the 15th International Conference on Wind Engineering; 2019 Sep 1–6; Beijing, China. p. 429–30

[28] Cheli F, Ripamonti F, Rocchi D, Tomasini G. Aerodynamic behaviour investigation of the new EMUV250 train to cross wind. Journal of Wind Engineering & Industrial Aerodynamics. 2010;**98**:189–201

[29] Premoli A, Rocchi D, Schito P, Tomasini G. Comparison between steady and moving railway vehicles subjected to crosswind by CFD analysis. Journal of Wind Engineering & Industrial Aerodynamics. 2016;**156**: 29–40

[30] Tominaga Y, Shirzadi M. RANS CFD modeling of the flow around a thin windbreak fence with various porosities: Validation using wind tunnel measurements. Journal of Wind Engineering & Industrial Aerodynamics. 2022;**230**:105176

[31] Niu J, Zhou D, Wang Y. Numerical comparison of aerodynamic performance of stationary and moving trains with or without windbreak wall under crosswind. Journal of Wind Engineering & Industrial Aerodynamics. 2018;**182**:1–15

[32] Niu J, Zhang Y, Chen Z, Li R, Yao H. Investigation of aerodynamic behaviour of a high-speed train on different railway infrastructure scenarios under crosswind. Wind Struct. 2022;**35**:405–418

[33] Deng E, Yang W, Lei M, Zhu Z, Zhang P. Aerodynamic loads and traffic safety of high-speed trains when passing through two windproof facilities under crosswind: A comparative study. Engineering Structures. 2019;**188**: 320–339

[34] Deng E, Yang W, He X, Zhu Z, Wang H, Wang Y, et al. Aerodynamic response of high-speed trains under crosswind in a bridge-tunnel section with or without a wind barrier. Journal of Wind Engineering & Industrial Aerodynamics. 2021;**210**:104502

[35] Xue R-D, Xiong X-H, Chen G, Li X-B, Liu B. Numerical comparison of aerodynamic performance between stationary and moving trains with varied-height windbreak wall under crosswind. Alexandria Engineering Journal. 2025;**110**:540–556

[36] Xu R, Wu F, Zhong M, Li X, and Ding J. Numerical investigation on the aerodynamics and dynamics of a high-speed train passing through a tornado-like vortex. Journal of Fluids and Structures. 2020;**96**:103042

[37] Diedrichs B, Ekequist M, Stichel S, Tengstrand H. Quasi-static modelling of wheel-rail reactions due to crosswind effects for various types of high-speed rolling stock. Proceedings of the Institution of Mechanical Engineers, Part F: Journal of Rail and Rapid Transit. 2004;**218**:133–148

[38] Liu D, Wang T, Liang X, Meng S, Zhong M, Lu Z. High-speed train overturning safety under varying wind speed conditions. Journal of Wind Engineering & Industrial Aerodynamics. 2020;**198**:104111

[39] Hibino Y, Shimomura T, Tanifuji K. Full-scale experiment on the behavior of a railway vehicle being subjected to lateral force. Journal of Mechanical Systems for Transportation and Logistics. 2010;**3**:35–43

[40] Misu Y, Ishihara T. Prediction of frequency distribution of strong crosswind in a control section for train operations by using onsite measurement

and numerical simulation. Journal of Wind Engineering & Industrial Aerodynamics. 2018;**174**:69–79

[41] Hibino Y, Kanemoto H, Shimomura T, Tanifuji K. Safety evaluation of railway vehicle against crosswind applying a full-vehicle simulation model. Transactions Of The Japan Society Of Mechanical Engineers Series C. 2010;**76**:3066–3076

[42] Sesma I, Vinolas J, San Emeterio A, Gimenez JG. A comparison of crosswind calculations using a full vehicle and a simplified 2D model. Proceedings of the Institution of Mechanical Engineers, Part F: Journal of Rail and Rapid Transit. 2012;**226**:305–317

[43] Thomas D, Diedrichs B, Berg M, Stichel S. Dynamics of a high-speed rail vehicle negotiating curves at unsteady crosswind. Proceedings of the Institution of Mechanical Engineers, Part F: Journal of Rail and Rapid Transit. 2010;**224**:567–579

[44] Zhang D, Ishihara T. Numerical study of tornado-induced unsteady crosswind response of railway vehicle using multibody dynamic simulations. Journal of Wind Engineering & Industrial Aerodynamics. 2022;**222**:104919

[45] Heleno R, Montenegro PA, Carvalho H, Ribeiro D, Calcada R, Baker CJ. Influence of the railway vehicle properties in the running safety against crosswinds. Journal of Wind Engineering & Industrial Aerodynamics. 2021;**217**:104732

[46] Burgers JM. A mathematical model illustrating the theory of turbulence. Advances in Applied Mechanics. 1948;**1**:171–199

[47] Rott N. On the viscous core of a line vortex. ZAMP Zeitschrift Für Angewandte Mathematik Und Physik. 1958;**9**:543–553

[48] Thomas D, Berg M, Stichel S. Measurements and simulations of rail vehicle dynamics with respect to overturning risk. Vehicle System Dynamics. 2010;**48**:97–112

[49] Brady SP, O'Brien EJ. Effect of vehicle velocity on the dynamic amplification of two vehicles crossing a simply supported bridge. Journal of Bridge Engineering. 2006;**11**:250–256

[50] Hibino Y, Kanemoto H, Sakuma Y. A study on vehicle response to a sudden gust of wind. Transactions Of The Japan Society Of Mechanical Engineers Series C. 2013;**79**:3410–3419

Section 2

Infrastructure Resilience and Intelligent Signaling

Chapter 3

Integrated Track Circuit Technology Based on All-Electronic Interlocking Systems in China

Haonan Feng, Yingjie Ji, Zhiguo Liang, Hongli Guo and Xiaojiao Ma

Abstract

This chapter presents a pioneering study on the development and implementation of integrated track circuit technology within the framework of all-electronic interlocking and unified train control systems in China. Addressing the critical issue of fragmentation in conventional railway signaling infrastructure, the research proposes a cohesive architectural solution that integrates disparate track circuit systems – such as the 25 Hz phase-sensitive track circuit for stations and the ZPW-2000 jointless track circuit for line sections – into a standardized and interoperable platform. Through experimental investigation and laboratory testing, the transmission characteristics of the integrated track circuit are evaluated under varying operational conditions, with emphasis on the influence of key parameters, including ballast resistance, rail length, and receiver type. The study further develops mathematical models to analyze transmission loss, impedance matching, and resilience to harmonic interference. Empirical results demonstrate that the integration of microelectronic receivers and customized impedance-matching components significantly enhances system reliability and anti-interference performance. The implementation of a domestically developed safety integrity level (SIL 4)-compliant electronic control module with a two-out-of-two architecture is also highlighted, marking a critical step toward the full electronization of interlocking systems. This work provides both theoretical and practical foundations for the design of next-generation integrated track circuits, supporting the ongoing digitalization and standardization of railway signaling systems in China.

Keywords: railway signaling systems, all-electronic interlocking, integrated track circuit, ZPW-2000, digitalization, standardization

1. Introduction

Track circuits are fundamental components in railway signaling systems, used to detect the occupancy status of railway tracks and facilitate communication between ground equipment and trains by transmitting movement authority information. Based on their application locations, track circuits are categorized into station track circuits and section track circuits. Station track circuits are deployed within station areas to monitor track occupancy without serving as block sections. Common types include 25 Hz phase-sensitive track circuits, JZXC-480 power-frequency AC continuous track circuits, 50 Hz phase-sensitive track circuits, high-voltage pulse track circuits, and ZPW-2000 integrated station track circuits.

Section track circuits are used in automatic block systems to monitor the occupancy of each block section and transmit movement authorities. The predominant type is the ZPW-2000 series jointless track circuit. On conventional and semi-high-speed railways, ZPW-2000 encoding is typically relay-based, while on high-speed lines, it employs communication-based encoding via Traffic Control Centers.

Globally, railway signaling systems are evolving toward digitalization and integration. Systems such as the European Train Control System (ETCS) and Positive Train Control (PTC) in the United States emphasize interoperability and continuous communication. China's approach, particularly through the ZPW-2000 series, offers a unique integration of station and section track circuits into a cohesive architecture, enhancing operational continuity and reliability.

2 Traditional track circuit in railway

2.1 A 25 Hz phase-sensitive track circuit in railway station

China's electrified railways widely use the 97-type 25 Hz phase-sensitive track circuit within stations. This system operates with a continuous 25 Hz AC power supply and employs an AC two-element track relay at the receiving end. Developed by the China Academy of Railway Sciences (CARS), the system operates as shown in **Figure 1**. A 25 Hz power supply panel provides both track and local power, with the local power leading the track power by 90°. The track power is transmitted via cables from the equipment room to the track, passing through a series of transformers and reactors before returning to energize the track relay. When the track is unoccupied, the relay remains energized; when occupied, the relay drops due to shunting.

However, the electromechanical relay has a low return coefficient (~50%) and is prone to contact jamming, which historically led to safety incidents. It is also susceptible to 50 Hz interference from electric locomotives. To address these issues, CARS developed the JXW-25B microelectronic receiver, which improves reliability and anti-interference performance. The receiver processes signals digitally and drives an execution relay based on phase and frequency criteria, as shown in **Figure 2**.

The microelectronic phase-sensitive track circuit receiver comprises four main components: an input unit, a computing unit, an output unit, and a power supply. The input circuit conditions the local signal via opto-isolation and analog-to-digital conversion, then relays it to the computing unit. The computing unit carries out digital signal processing, while the output unit drives the track execution relay.

DOI: http://dx.doi.org/10.5772/intechopen.1013126

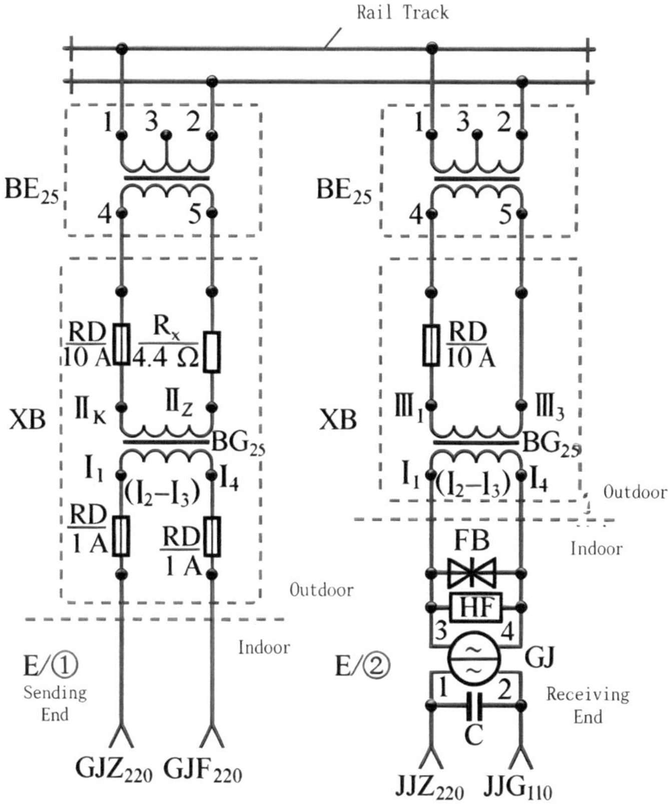

Figure 1.
Schematic diagram of a 25 Hz phase-sensitive track circuit.

2.2 ZPW-2000 with jointless track circuit in railway section

The ZPW-2000 jointless track circuit used in railway sections may be divided into two primary categories: relay-based encoding and communication-based encoding. Relay-based encoding is predominantly employed in conventional speed lines, where encoding conditions are supplied by relay circuits. In this configuration, the transmitter adopts either an N + 1 or a 1 + 1 primary/backup redundancy arrangement, while the receiver operates in a 0.5 + 0.5 parallel redundancy mode.

Communication-based encoding, on the contrary, is mainly utilized in passenger-dedicated lines and high-speed railways. In this system, encoding conditions are provided by the Traffic Control Center through communication interface boards using specific communication protocols. The transmitter employs a 1 + 1 primary/backup redundancy mode, and the receiver similarly uses a 0.5 + 0.5 parallel redundancy setup, while also transmitting track status information via communication links.

Outdoor equipment establishes an electrical insulation joint through the use of matching transformers, tuning units, and air-core coils. The structural design of the ZPW-2000 jointless track circuit in railway sections includes two main types: electrical–electrical insulation and electrical–mechanical insulation, as illustrated in **Figures 3** and **4**.

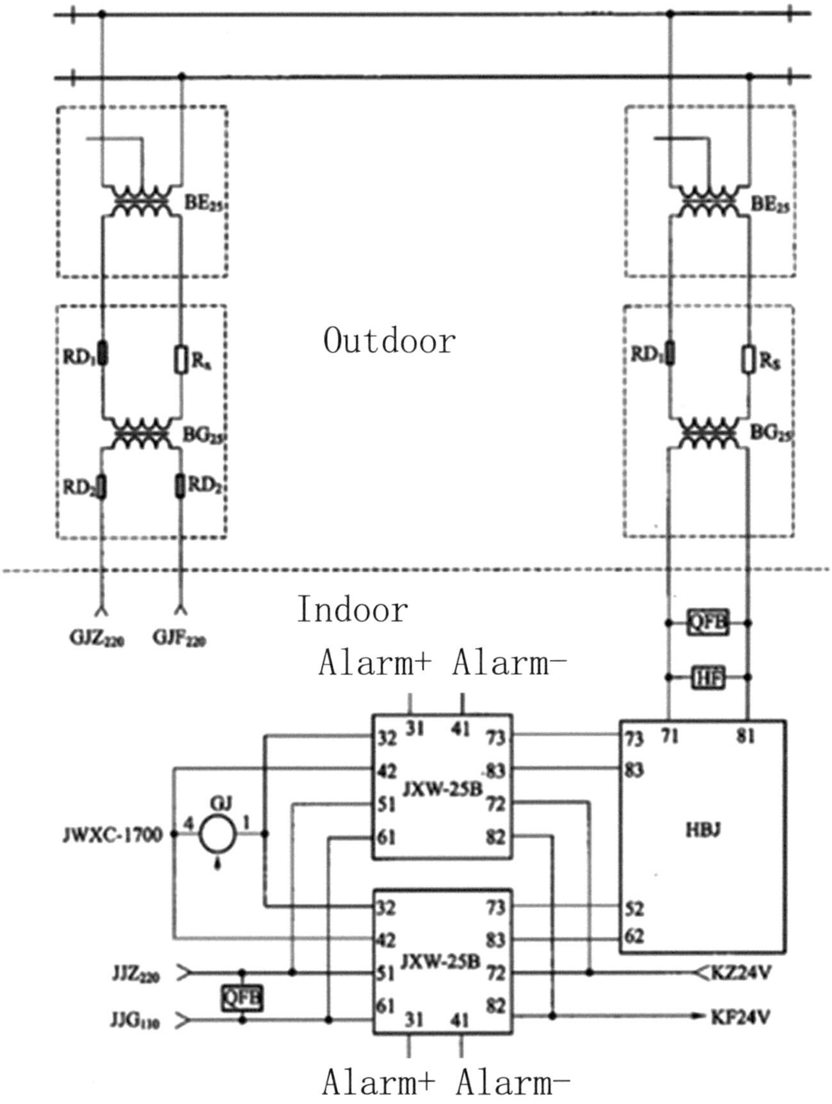

Figure 2.
JXW-25B microelectronic phase-sensitive track circuit principle.

3. Innovation on track circuits in China

3.1 integrated track circuit in railway station

To overcome the limitations associated with traditional track circuits, an integrated approach has been developed. To enable trains to continuously receive train operation control information from the ground, newly constructed lines, as well as some retrofitted existing lines, have adopted ZPW-2000 track circuits, including within station areas – a solution referred to as the integrated track circuit system. However, when the ZPW-2000 track circuit is deployed in station areas, the specific operational characteristics of station tracks necessitate that the tuning zone

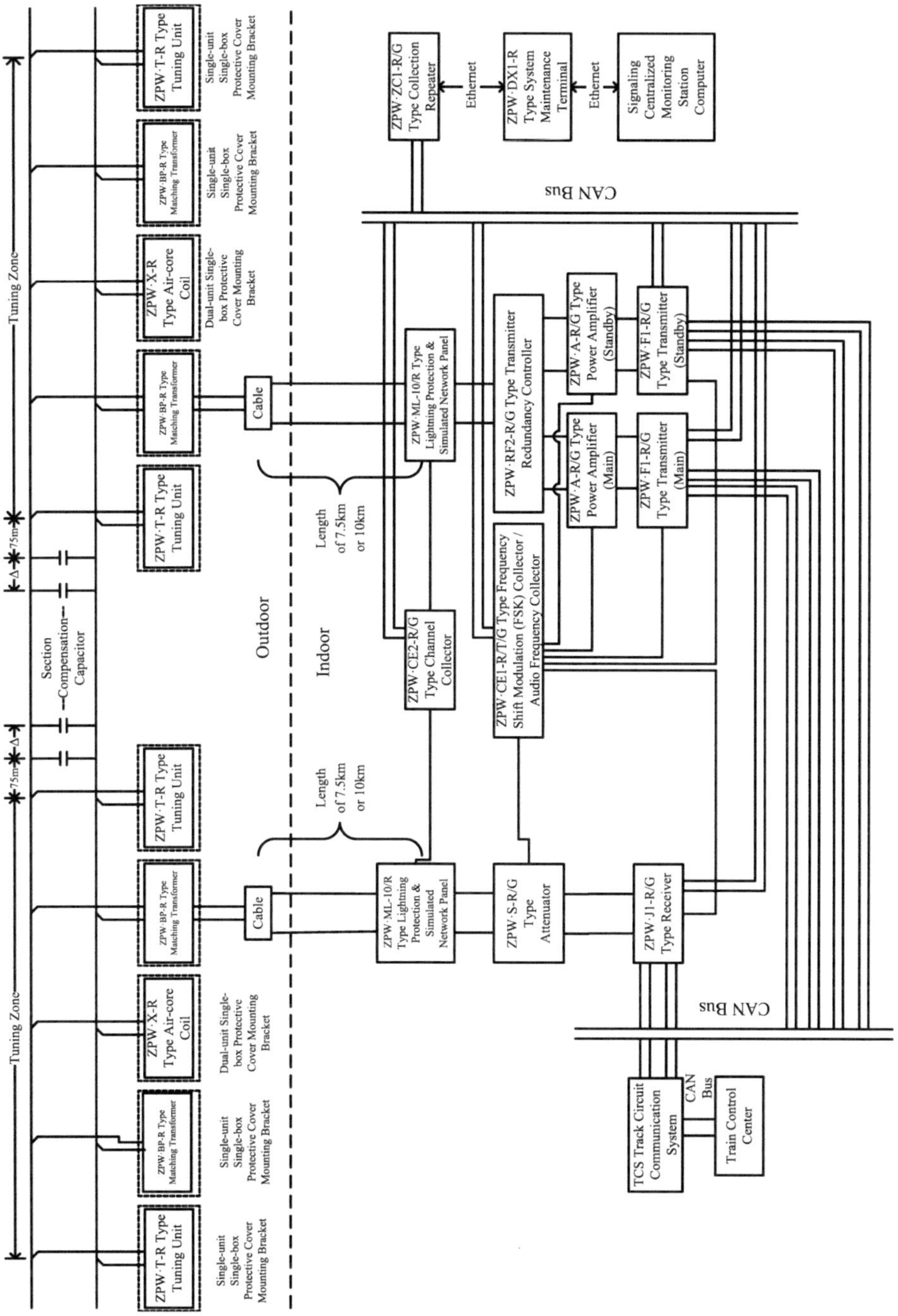

Figure 3.
Structure of the ZPW-2000 jointless track circuit with electrical–electrical insulation joints in railway sections.

be adapted to a mechanical insulation joint-based configuration. Mechanical insulation joints are available in four models: 1700, 2000, 2300, and 2600. Each mechanical insulation joint comprises a mechanical insulation joint air-core coil (SVA) connected in parallel with a corresponding tuning unit. Critically, the tuning unit must be selected such that it exhibits zero impedance at the signal carrier

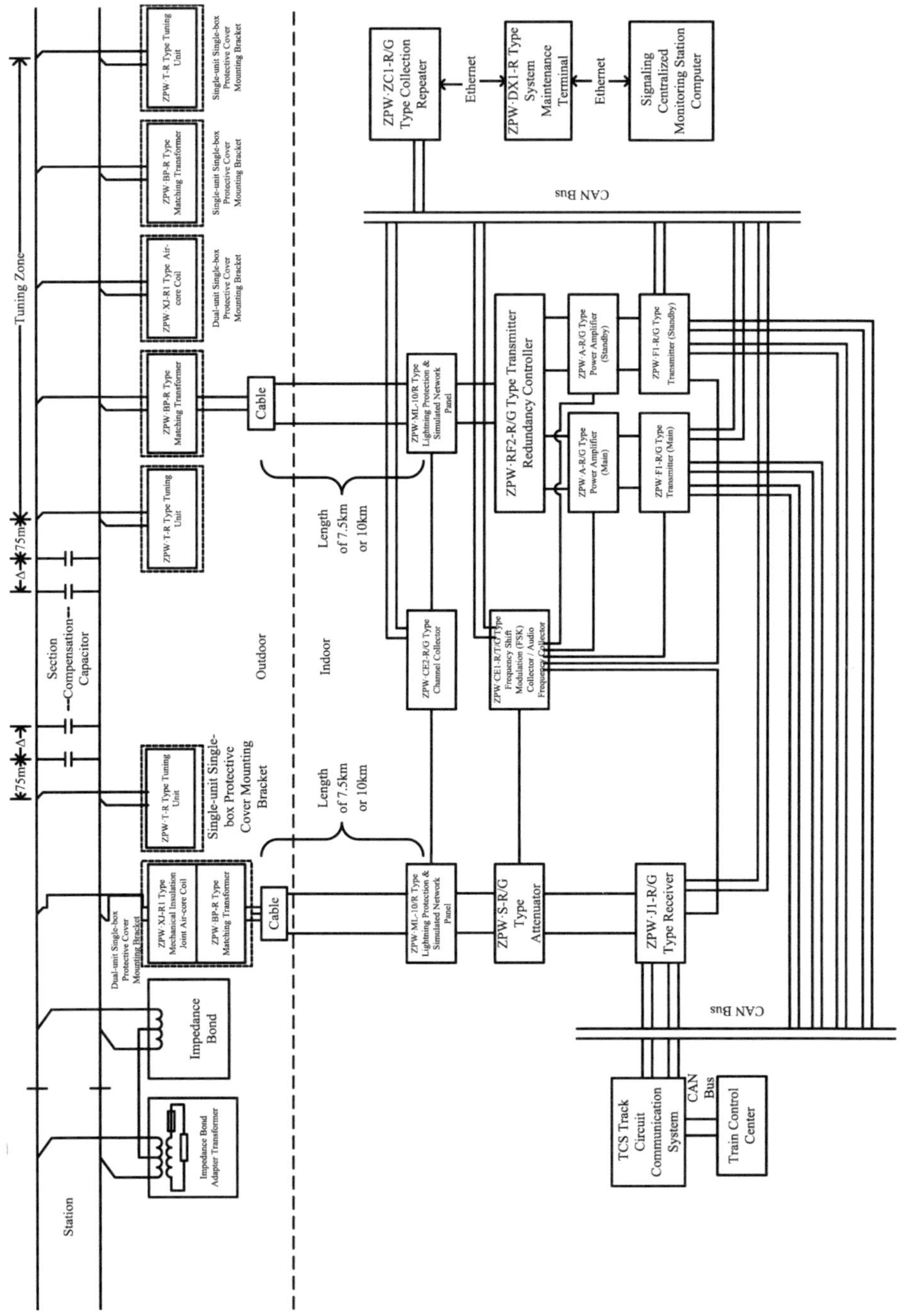

Figure 4.
Structure of the ZPW-2000 jointless track circuit with electrical–mechanical insulation joints in the railway.

frequency of the home section track circuit, thereby establishing a reliable transmission channel for the signals of that section.

To ensure that the ZPW-2000 integrated station track circuit achieves a transmission length comparable to that of the jointless ZPW-2000 track circuit used in line sections, the impedance value of the SVA is specifically designed based on the composite impedance of the 29-m tuning zone across each of the four carrier frequencies. The matching transformers used in stations differ from those employed in line sections primarily in their matching ratios. Furthermore, a choke transformer equipped with an adapter is utilized between adjacent track sections within the station track circuit. This arrangement provides a dedicated path for traction return current and helps mitigate interference caused by unbalanced traction currents, thereby enhancing the overall anti-interference capability of the track circuit. As indicated in **Figures 3** and **4**, the transmission cable length in existing jointless ZPW-2000 track circuits is typically 10 km, whereas in passenger-dedicated lines, this is reduced to 7.5 km.

Currently, high-speed railway lines utilize the same type of track circuit across both stations and sections, known as the ZPW-2000 integrated track circuit. The basic structure of the ZPW-2000 integrated track circuit used in stations is largely similar to that of the jointless ZPW-2000 track circuit deployed in line sections. Key differences include the use of mechanical insulation joints in the station configuration, the parallel connection of non-energized branches in turnout areas via turnout jumpers, and minor variations in the models of matching transformers. The indoor equipment, however, remains completely identical between the two applications.

3.2 Track circuit electronic control module

All-electronic interlocking systems offer advantages such as highly integrated electronics, simplified system architecture, high intelligence, robust functionality, flexible configuration, and extensive control coverage. These systems represent a growing trend in rail transit signaling systems worldwide [1]. In conventional computer-based interlocking systems, the execution layer that interfaces with trackside equipment typically employs safety relays and associated relay circuits. These are now being replaced by intelligent electronic execution units, which take over the functions of digital acquisition, drive execution circuits, and safety relay interfaces. Such units enable direct control of trackside devices – including switch machines and signal lights – thereby supporting interface expansion and facilitating the full electronization of interlocking systems [2].

While all-electronic interlocking systems have been widely implemented internationally, their adoption in China faces specific challenges [3]. Significant differences in the electrical characteristics of domestic signaling equipment, along with unique operational requirements, hinder the direct integration of foreign systems into China's rail transit control infrastructure [4]. As a result, there is a pressing need for independent development of all-electronic interlocking systems tailored to local conditions. Domestic signaling suppliers, including research institutions such as Lanzhou Jiaotong University, have developed computer-based interlocking electronic execution modules [5]. Although these systems are capable of communicating with computer-based interlocking platforms, their deployment within China's railways remains limited thus far [6].

Track circuits are essential signaling components in railways, but their reliance on complex relay circuits often leads to intricate wiring, elevated failure rates, and limited anti-interference performance [7]. Current domestic efforts to enhance track circuit performance still prioritize improving the anti-interference capabilities of receiving equipment. Nevertheless, the continued use of relay-based interfaces hinders the evolution of interlocking systems toward communication-based railway signaling architectures and is incompatible with the development of intelligent interlocking systems. Electronic control modules present a viable solution to these limitations. To enable real-time and secure acquisition of track circuit parameters and accurate determination of track section status (occupied or clear), these modules must demonstrate high levels of reliability and safety [8].

The independently developed electronic control unit for track circuits employs a two-out-of-two (2oo2) architecture. Following extensive testing and validation, it has been integrated into domestically produced all-electronic interlocking systems and successfully deployed in several rail transit projects across China, confirming both the reliability and practicality of the electronic control module.

The track circuit electronic control module comprises several key components: primary and backup track front-end boards, track rear-end boards, a universal motherboard, and an isolation matching unit. The track front-end board processes both track and local signals, determines track occupancy status, and communicates with the object controller to exchange information with the interlocking logic section. Upon receiving a section call command from the logic section, it returns the current section status. The universal motherboard reads address jumper settings to assign a unique identifier to each electronic control module. The isolation matching unit introduces the track signal and ensures dual-channel isolation and impedance matching. The track rear-end board receives the isolated and matched track signals from the isolation matching unit, along with local signal inputs, and interfaces with both the primary and backup front-end boards. A structural diagram of the track circuit electronic control module is provided in **Figure 5**.

The software for the track circuit electronic control module is designed in accordance with the EN 50129 standard. From the perspective of functional requirements and software architecture, the application software for the 25 Hz track

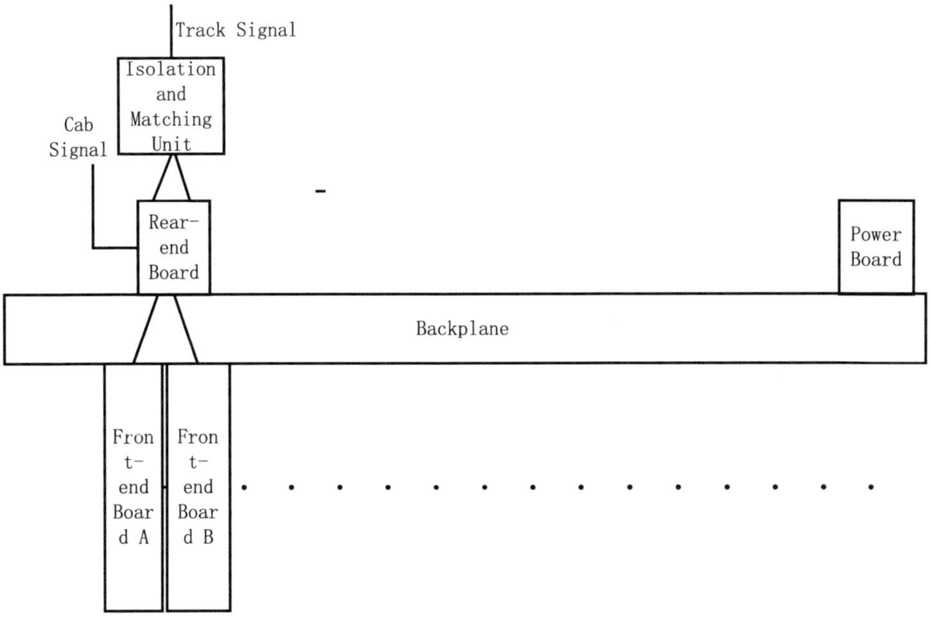

Figure 5.
Structural diagram of the track circuit electronic control module.

Name	Quantity	Failure rate(10^{-6}/h)	MTBF (h)
Power board	1	0.071102	14,064,320
Rear-end board	1	0.130837	76,43,098
Isolation and matching unit	1	0.310064	3,225,143
Front-end board	2	6.579121	151,996
Backplane	1	0.138317	7,229,786

Table 1.
Reliability parameters of components for the track circuit electronic control module.

circuit subsystem is divided into four modules: the application program, communication control, initialization, and driver components. Each of these is further implemented through multiple submodules. The software incorporates methods and technologies such as fault detection and diagnosis, error detection codes, fully defined interfaces, and structured design approaches to satisfy the SIL 4 (safety integrity level) design requirements.

In the design of the track circuit electronic control module, the FN^2 model was adopted for reliability analysis and development. This model references the MIL-HDBK-217 standard, published by the Reliability Analysis Center of the U.S. Department of Defense. The reliability calculation accounts for influencing factors, including product design characteristics, environmental and operational load effects, definitions of product failure, and random variables. It provides a comprehensive reliability prediction based on a physics-of-failure reliability model.

Reliability analysis was performed based on the structural diagram of the track circuit electronic control module (see **Figure 5**) and the reliability parameters of its components (see **Table 1**). The model assumed an indoor operating environment, a 100% system duty cycle, and an operating temperature of 40°C. The calculated failure rate of the electronic control module is $\lambda = 0.578249 \times 10^{-6}$/h, with a mean time between failures (MTBF) of 1,729,659 hours, meeting the SIL 4 design requirements.

4 Interference analysis of the track circuit

The ZPW-2000 jointless track circuit is widely adopted due to its strong immunity to interference, broad applicability, and ability to ensure stable train operation. As a critical component of the train control system, interference issues in adjacent sections of the track circuit occasionally occur. Thus, understanding interference is critical for the reliable operation of integrated track circuits [9].

4.1 Analysis of interference from the adjacent section

Adjacent section interference occurs when a train is operating within a given track circuit section, and signals from an adjacent track circuit cross the electrical insulation joint, interfering with the signals received by the train in the current section. If the amplitude of the interfering signal reaches the reception threshold of the onboard equipment, the device may simultaneously receive two different signal frequencies. This can compromise the accuracy of signal demodulation and potentially lead to dangerous malfunctions, thereby posing significant safety risks.

The causes of adjacent section interference are diverse and can generally be classified into two categories. The first category pertains to non-standard construction practices and equipment quality issues. These may include discrepancies in the calibration and adjustment of track circuits relative to specifications, problems with cable quality or improper installation, and excessive zero-impedance values. Such issues can be prevented through rigorous inspection during the construction phase. The second category involves failures that develop during the long-term operation of the equipment, such as faults within the tuning zone.

The ZPW-2000 jointless track circuit comprises two main parts: the main track circuit and the sub-track circuit within the tuning zone. The sub-track circuit is responsible for providing signal isolation between adjacent track circuits. However, faults in the tuning zone can compromise this isolation, leading to interference between adjacent sections and impairing the safety and efficiency of train operations. Therefore, research on adjacent section interference is of considerable importance. The underlying principle of this interference is illustrated in **Figure 6**.

During train operation, the train control receiver (TCR) equipment cannot selectively identify the exact carrier frequency of the current track circuit section. Instead, depending on the direction of travel, it receives all possible frequency signals associated with that direction. If a fault exists in the tuning zone, signals from an adjacent track circuit (e.g., Track Circuit 3) may leak across the tuning zone into Track Circuit 1. When the amplitude of this interfering signal surpasses a certain threshold, it adversely affects the normal operation of the TCR equipment.

4.2 Analysis of interference from adjacent lines

The ZPW-2000 track circuit adjacent-line interference occurs in multi-track parallel sections when the receiver of the local track circuit – or the onboard equipment of a train on the local line – receives interfering signals from an adjacent track circuit operating in the same direction. In the subsequent analysis, the affected

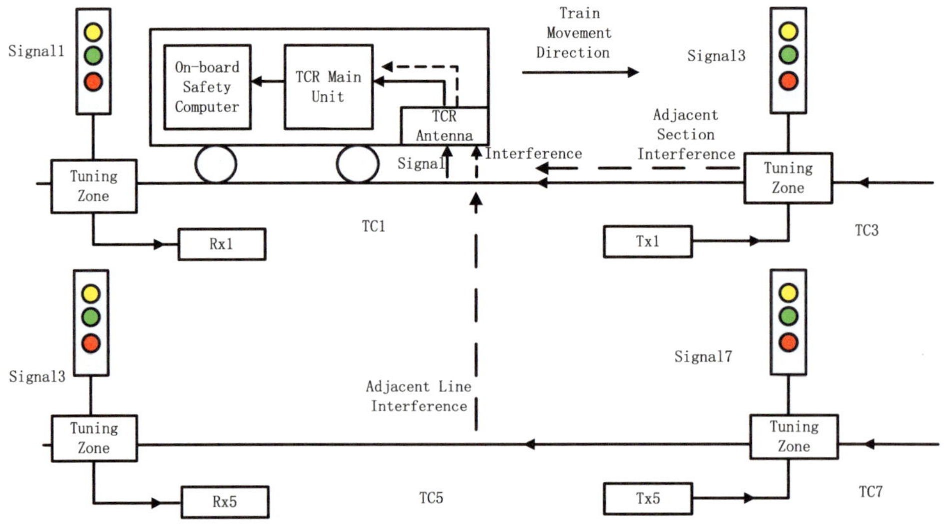

Figure 6.
ZPW-2000 track circuit interference schematic diagram.

line is termed the disturbed circuit, and the adjacent line is referred to as the disturbing circuit. The mechanism of adjacent-line interference is illustrated in **Figure 6**.

As shown in **Figure 6**, assume Track Circuit 5 is in the adjusted state, while Track Circuit 1 is in the shunted state. If adjacent-line interference from Track Circuit 5 affects Track Circuit 1, two potential consequences may arise:

(1) The onboard equipment may receive signals from Track Circuit 1 alongside interference from Track Circuit 5. If the amplitude of the interfering signal exceeds a certain threshold, the cab signaling equipment may generate incorrect movement authorities, ultimately reducing operational efficiency and potentially leading to accidents.

(2) A relatively high-amplitude signal current from the disturbing circuit may flow into Receiver 1 of Track Circuit 1 (which is in the shunted state), implying an insufficient shunting effect from the train wheelsets. As a result, the track relay of Track Circuit 1 may erroneously pick up, causing Signal Light 1 to change from red to green. A following train may then misinterpret Track Circuit 1 as unoccupied and enter the section, creating a risk of collision or other serious safety incidents.

The primary causes of adjacent-line interference include crosstalk between transmission cables and coupling interference. Coupling interference comprises inductive coupling between rails, capacitive coupling, leakage conduction through ballast resistance, and electromagnetic field coupling. Since ZPW-2000 track circuits generally use SPT digital cables for signal transmission, cable crosstalk can largely be avoided through strict adherence to application specifications. Therefore, coupling interference remains the predominant factor contributing to adjacent-line interference.

5 Performance analysis of the track circuit

5.1 Simulation environment of 25 Hz phase-sensitive rail circuit

To investigate the transmission characteristics of existing 25 Hz phase-sensitive track circuit equipment, an experimental setup was constructed to simulate a 25 Hz track circuit. The configuration includes a 25 Hz power supply, simulated transmission-end transformer units, impedance bonds, rail simulation units, simulated reception-end transformer units, binary two-position relays, microelectronic receivers, and other associated components. The overall structure of the test setup is depicted in **Figure 7**.

To facilitate debugging and testing, the majority of the equipment components are integrated into a standard equipment rack, as illustrated in **Figure 8**. Both the relay receivers and microelectronic receivers utilize a shared signal transmission path, with the option to switch between them via terminals located at the rear of the rack.

5.2 Simulation environment of ZPW-2000 rail circuit

To investigate the transmission characteristics of existing ZPW-2000 track circuit equipment, an experimental test platform was constructed. The setup

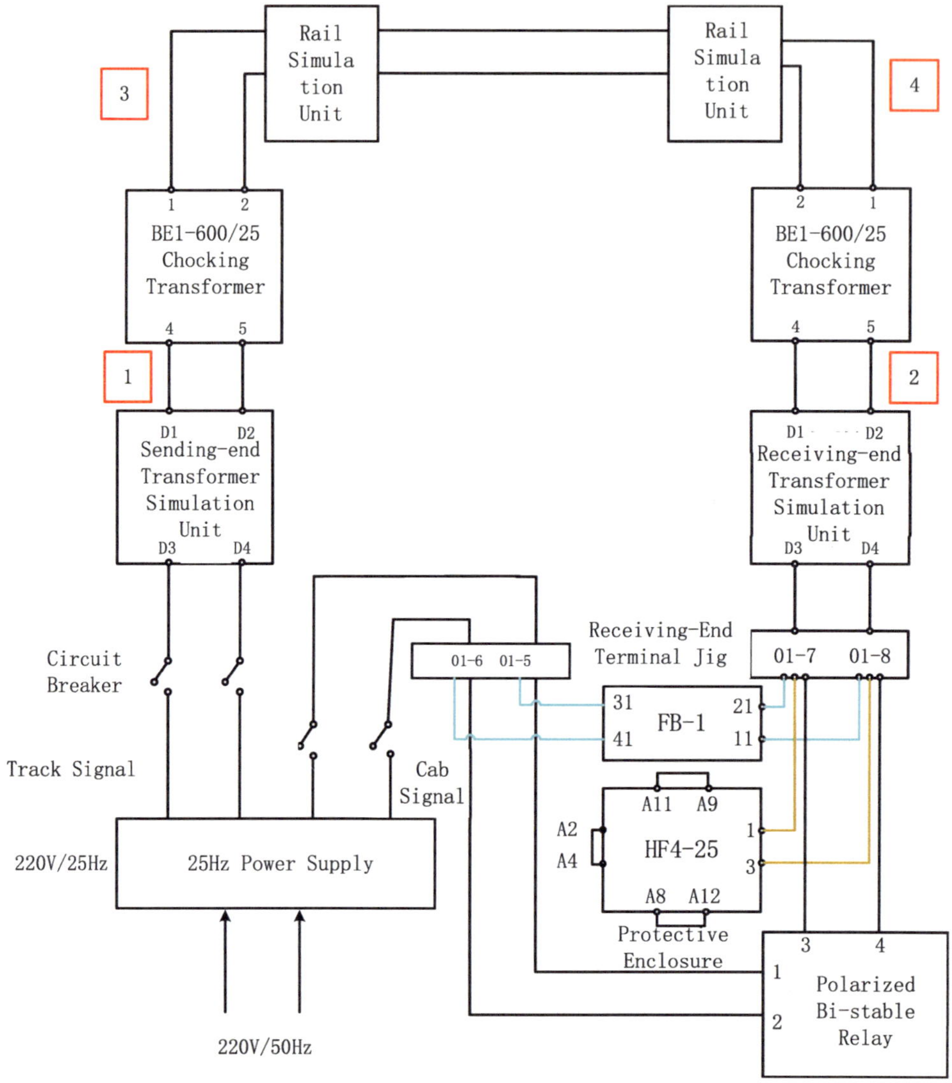

Figure 7.
Structure of the 25 Hz track circuit test platform.

comprises two frequency-shift cabinets, two interface cabinets, outdoor equipment, and rail simulation units emulating four line sections, as well as corresponding outdoor equipment and rail simulation units for one station section. A photograph of the actual test environment is provided in **Figure 9**.

5.3 Mathematical model on track circuit

A simplified diagram of the track circuit structure is provided in **Figure 10**. This chapter addresses the modeling principles and simulation methods for the track four-terminal network, analyzes its various operating states, and finally presents a modular simulation process for the network [10].

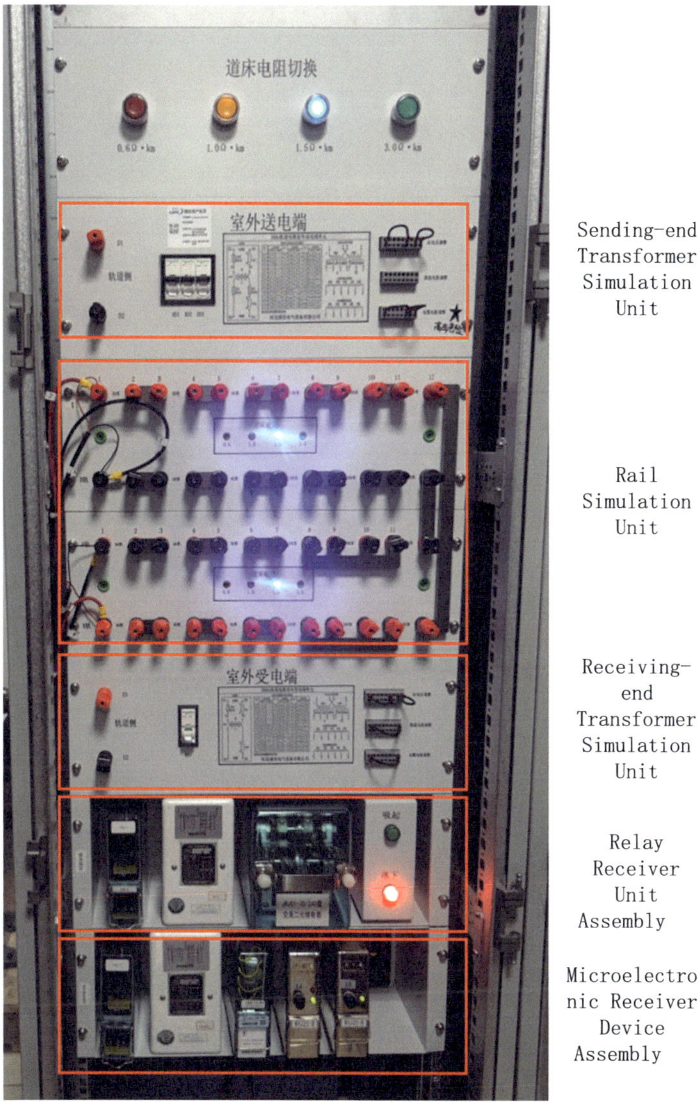

Figure 8.
A 25 Hz track circuit test bench cabinet.

The two rails situated between the transmitting-end and receiving-end electrical insulation joints function as conductors, carrying the output Frequency Shift Keying (FSK) signal from the transmitter to the input of the receiver, thereby forming a track four-terminal network for signal transmission. Due to the inherent resistance and parasitic inductance of the rails, as well as the leakage resistance and parasitic capacitance (leakage capacitance) between them, the audio FSK signal undergoes attenuation during transmission. Environmental factors such as temperature and humidity – which affect the ballast and sleeper conditions – can cause significant variations in leakage resistance and capacitance, thereby influencing the extent of signal attenuation. It is, therefore, essential to study the impedance characteristics,

Figure 9.
Physical photo of the ZPW-2000 track circuit test environment.

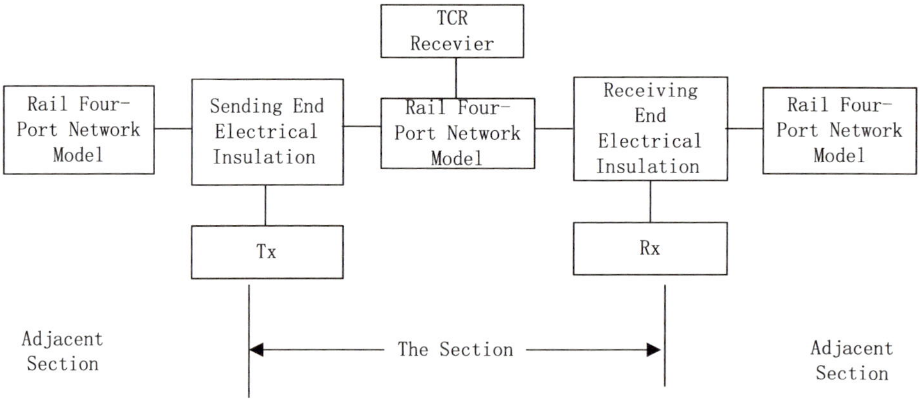

Figure 10.
Track circuit structure diagram.

voltage and current distributions along the track four-terminal network, and the factors that influence them.

The track four-terminal network operates in three primary states: adjusted, shunted, and broken rail. The adjusted state corresponds to an unoccupied track section with no train present. When a train is passing over the section, the network enters the shunted state, which alters the network's structure and modifies the voltage and current distribution along the line, as well as the signal attenuation characteristics. A broken rail represents a fault condition in which the rail is physically fractured. This section focuses on the mathematical modeling and simulation of the track four-terminal network in both the adjusted and shunted states, including both single and double shunting scenarios.

The track circuit continuously provides vital information such as section occupancy, broken rail detection, train positioning, and movement authorities to the zone control center and onboard ATP systems. Essentially, it acts as an electrical pathway – comprising two parallel conductors – for transmitting voltage and current signals. The physical process of signal propagation along this path can be analyzed either from an electromagnetic field perspective or a circuit theory perspective.

Given the fewer variables and simpler relational expressions involved, the circuit method is generally preferred. By exploiting the correspondence between electromagnetic fields and circuits under sinusoidal steady-state conditions, the field problem can be transformed into an equivalent circuit problem, simplifying analysis and improving intuitive understanding.

To analyze voltage and current transmission using circuit theory, the transmission line must be treated as a distributed parameter circuit. In such circuits, voltage and current are functions of both time and spatial position, meaning the analysis must account for their variation along the length of the line. Since real-world conductors and insulation are non-ideal, the model must include distributed resistance R (representing transverse power loss) and conductance G (representing longitudinal leakage), as well as capacitance C and inductance L arising from the electric and magnetic fields between the rails. For long transmission lines, such as those in track circuits, the finite propagation time of electromagnetic waves cannot be neglected. Thus, using the lumped parameter approximation, the line is divided into differential segments of length dx. Within each segment, the parameters R, L, C, and G are considered lumped, and voltage and current are treated as functions of time only. The complete transmission line is then modeled as an infinite series of these discrete segments. Given a sinusoidal excitation source, the resulting voltage and current responses will also be sinusoidal at the same frequency.

The track circuit is a uniformly distributed parameter circuit. The rail resistance R, inductance L, inter-rail leakage capacitance C, and leakage conductance G are uniformly distributed along the line and are typically expressed per unit length (per kilometer). Moreover, since the two rails maintain consistent geometry, relative positioning, and surrounding medium throughout their length, the track four-terminal network can be accurately analyzed using uniform transmission line theory. In practical track circuits, transient signal components decay rapidly, allowing the analysis to focus solely on the steady-state solution. Under sinusoidal steady-state conditions, the second-order partial differential equations describing the transmission line can be reduced to first-order differential equations.

The equivalent circuit of the track four-terminal network is illustrated in **Figure 10**. According to uniform transmission line theory, the entire network can be modeled as an infinite series of differential segments. The parameters for a differential segment of length dx are defined as follows [11]:

$$Z = R_g + j\omega L_g \tag{1}$$

$$G = 1/R_o + j\omega C_o \tag{2}$$

where R_g is the rail resistance per unit length; R_o is the rail leakage resistance per unit length; L_g is the rail inductor per unit length; C_o is the rail leakage capacitance per unit length. $\omega = 2\pi f$, where f is the frequency at which signals are transmitted on the rail.

According to Kirchhoff's laws, the voltage–current relationship across any differential segment dx can be described by the following system of differential equations and their analytical solutions:

$$\begin{cases} \dfrac{dU}{dx} = ZI \\ \\ \dfrac{dI}{dx} = -GU \end{cases} \tag{3}$$

$$\begin{bmatrix} U \\ I \end{bmatrix} = \begin{bmatrix} \frac{1}{2}\left(e^{\sqrt{GZ}x} + e^{-\sqrt{GZ}x}\right)U_0 + \frac{1}{2}\sqrt{\frac{Z}{G}}\left(e^{\sqrt{GZ}x} - e^{-\sqrt{GZ}x}\right)I_0 \\ \frac{1}{2}\left(e^{\sqrt{GZ}x} - e^{-\sqrt{GZ}x}\right)U_0 + \frac{1}{2}\left(e^{\sqrt{GZ}x} + e^{-\sqrt{GZ}x}\right)I_0 \end{bmatrix} \tag{4}$$

where

U is the voltage distribution along the track four-terminal network, I is the current distribution along the track four-terminal network, U_0 is the voltage across the equivalent load of the track four-terminal network, I_0 is the current through the equivalent load of the track four-terminal network, and x is the distance from any point along the track four-terminal network to the receiving end.

5.3.1 Adjusted state

The output voltage signal from the transmitting-end electrical insulation joint is transmitted through the rails and applied to the receiving-end electrical insulation joint. It is generally assumed that the input voltage and load resistance of the track four-terminal network are known, as solving circuit problems typically begins with these given variables to derive other responses. Based on this, the voltage, current, and input impedance at any point along the track four-terminal network can be calculated. The equivalent circuit is shown in **Figure 11**.

Assuming the length of the track circuit is L and the distance from any point along the rail surface to the end of the track is denoted as x.

When $x = L$, $U_{ini} = U_L$, and $\frac{U_0}{I_0} = R_{end}$, we can get

$$U_0 = \frac{U_{ini}}{\frac{1}{2}A_1 + \frac{1}{2}\sqrt{\frac{Z}{G}\frac{B_l}{R_{end}}}} \tag{5}$$

By applying the aforementioned formulae, the voltage and current distribution along the track, as well as the input resistance from any point on the track to the receiving end, can be derived:

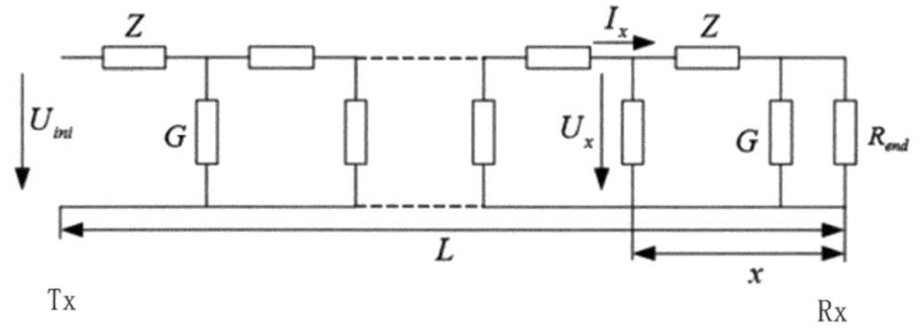

Figure 11.
Equivalent circuit of the track four-terminal network in an adjusted state.

$$U_x = \frac{AR_{end} + \sqrt{\frac{Z}{G}}B}{A_1R_{end} + \sqrt{\frac{Z}{G}}B_l}U_{ini} \tag{6}$$

$$I_x = \frac{A + \sqrt{\frac{G}{Z}}BR_{end}}{A_1R_{end} + \sqrt{\frac{Z}{G}}B_l}U_{ini} \tag{7}$$

$$R_{inx} = \frac{AR_{end} + \sqrt{\frac{Z}{G}}B}{A + \sqrt{\frac{G}{Z}}BR_{end}} \tag{8}$$

where:
$A = e\sqrt{GZx} + e - \sqrt{GZx}$, $B = e^{\sqrt{GZx}} - e^{-\sqrt{GZx}}$
A_l, B_l are A, B values at $x = L$;
x is the distance from any point along the track to the receiving end; L is the length of the track four-terminal network; U_x is the voltage at x meters from the receiving end; I_x is the current at x meters from the receiving end; and R_{inx} is the input impedance of the track four-terminal network at x meters from the receiving end.

5.3.2 Shunted state

Under shunted conditions, the shunting resistance is composed of the train axle resistance, wheel resistance, and the contact resistance between the wheels and the rails. Its value typically ranges from 0.04 to 0.15 ohms. Since the shunting resistance is significantly smaller than the impedance of the subsequent track four-terminal network, it is assumed that, at the shunting point, the input impedance of the track four-terminal network equals the shunting resistance, as illustrated in **Figure 12**.

Let the distance from the shunting point to the receiving end be L_2, and the shunting resistance be R_f. By applying the aforementioned formulae, the voltage distribution, current distribution, and input impedance along the track between the transmitting end and the shunting point can be derived:

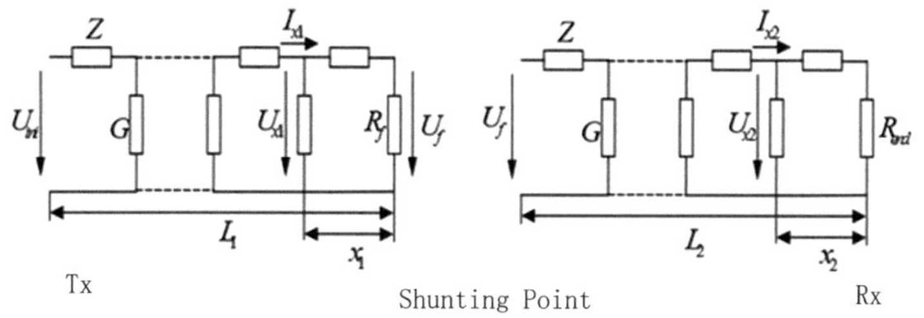

Figure 12.
Equivalent circuit of the track four-terminal network under single-shunted condition.

$$U_{x1} = \frac{AR_f + \sqrt{\frac{Z}{G}}B}{A_{L1}R_f + \sqrt{\frac{Z}{G}}B_{L1}} U_{ini} \tag{9}$$

$$I_{x1} = \frac{A + \sqrt{\frac{G}{Z}}BR_f}{A_{L1}R_f + \sqrt{\frac{Z}{G}}B_{L1}} U_{ini} \tag{10}$$

$$R_{inx1} = \frac{AR_f + \sqrt{\frac{Z}{G}}B}{A + \sqrt{\frac{G}{Z}}BR_f} \tag{11}$$

The voltage across the terminals of the divider resistor is

$$U_f = \frac{2R_f}{A_{L1}R_f + \sqrt{\frac{Z}{G}}B_{L1}} U_{ini} \tag{12}$$

where

A_{L1} and B_{L1} are the corresponding values of A and B when $x = L_1 = L - L_2$.

The distribution of voltage, current, and input impedance along the track between the junction point and the receiving end can be derived as:

$$U_{x2} = \frac{AR_{end} + \sqrt{\frac{Z}{G}}B}{A_{L2}R_{end} + \sqrt{\frac{Z}{G}}B_{L2}} U_f \tag{13}$$

$$I_{x2} = \frac{A + \sqrt{\frac{G}{Z}}BR_{end}}{A_{L2}R_{end} + \sqrt{\frac{Z}{G}}B_{L2}} U_f \tag{14}$$

$$R_{inx2} = \frac{AR_{end} + \sqrt{\frac{Z}{G}}B}{A + \sqrt{\frac{G}{Z}}BR_{end}} \tag{15}$$

All these electrical quantities are complex numbers, and the formulae employ complex number operations. The resistance value from the shunting point to the receiving end decreases as L_2 decreases, and may become comparable to R_f. It is also necessary to discuss the voltage and current distribution on the segment from the shunting point L_2 to the receiving end. When L_2 is relatively large, the current on this segment is very small.

5.4 Simulation

Using the mathematical model described in Section 5.3, we simulated the track circuit under different conditions. The simulation parameters for the track circuit are shown in **Table 2** [12].

Physical quantity parameter	Value
Input voltage (V)	7.47
Rail resistance (Ω/km)	6.5
Rail inductance (H/km)	1.35×10^{-3}
Physical quantity parameter (Ω/km)	1.5
Rail leakage capacitance (F/km)	1.2×10^{-6}
Rail length (km)	1
Carrier frequency (kHz)	1.7
Load impedance at the four ends of the track network (Ω)	2.7
Distance from the junction point to the receiving end (km)	0.1
Divergence resistor (Ω)	0.15

Table 2.
Simulation parameters table.

According to the simulation parameters in **Table 2** above, the simulation results are shown in **Figures 13** to **15**.

As demonstrated by the simulation results in **Figures 13–15**, the impedance of the track four-terminal network increases at higher carrier signal frequencies, resulting in increased signal attenuation along the rails for a constant input voltage. Both voltage and current decrease steadily from the transmitting end to the receiving end. The shape of these attenuation curves is influenced by ballast bed conditions: under high ballast bed resistance, low rail leakage results in an

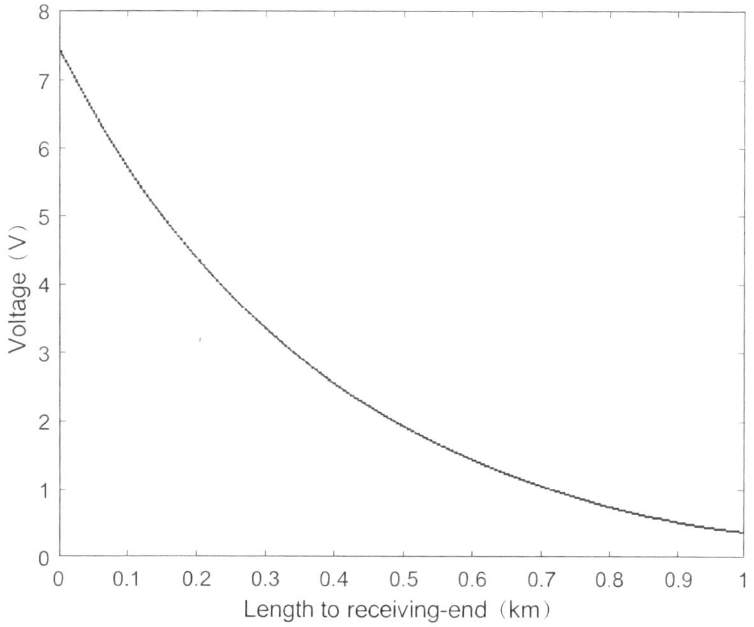

Figure 13.
Voltage distribution along the track circuit in an adjusted state.

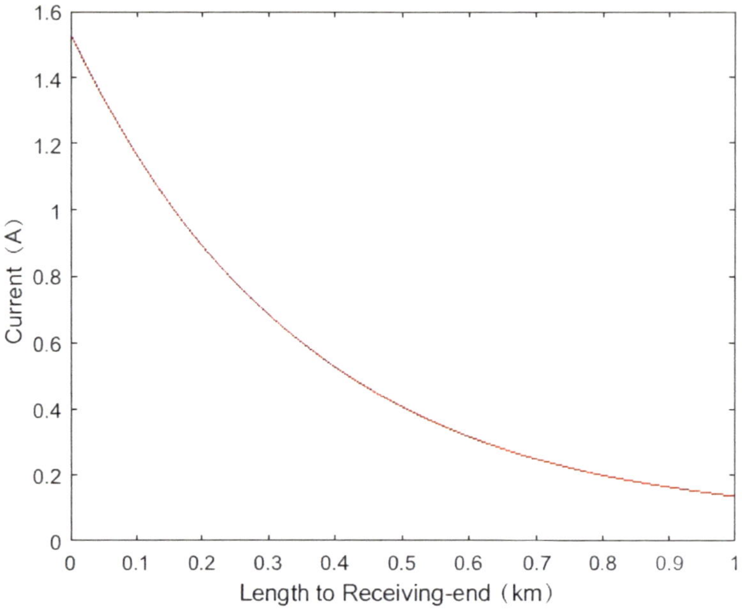

Figure 14.
Current distribution along the track circuit in an adjusted state.

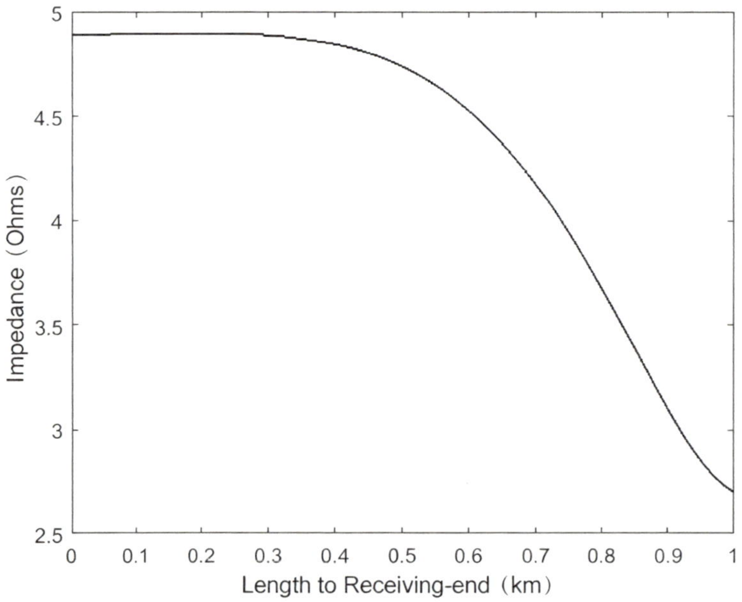

Figure 15.
Impedance distribution along the track circuit in an adjusted state.

approximately linear decline dominated by rail impedance, while under low ballast bed resistance, greater leakage current introduces curvature into the attenuation profile. The ratio of the receiving-end voltage to the transmitting-end

DOI: http://dx.doi.org/10.5772/intechopen.1013126

voltage is measured at 0.05. Additionally, the voltage attenuation rate increases with longer track lengths and is further influenced by ballast bed parameters – specifically, lower rail leakage capacitance leads to higher leakage currents and greater attenuation, which can be compensated by adding external capacitance between the rails.

As shown in the simulation results in **Figures 16–18**, the voltage along the four-terminal network exhibits a rapid drop from the transmitting end to the shunting point. This is due to the large shunting current, which results in a substantially higher voltage drop across the rail compared to that caused by leakage current, leading to an approximately linear decrease. Beyond the shunting point toward the receiving end, the voltage continues to diminish following the same attenuation pattern observed in the adjusted state, albeit from an already reduced level. The voltage ratios are measured at 0.09 from the transmitting end to the shunting point and 0.06 from the shunting point to the receiving end, resulting in an overall end-to-end ratio of 0.005. This indicates a tenfold increase in attenuation compared to the adjusted state. The significantly higher voltage attenuation ratio during shunting produces a very low output voltage, enabling the receiver to reliably detect track occupancy. Meanwhile, the relatively large shunting current passing through the wheelsets can be readily detected by onboard equipment. Furthermore, the magnitude of the shunting current is influenced by the location of the shunting point, decreasing as the point moves closer to the receiving end.

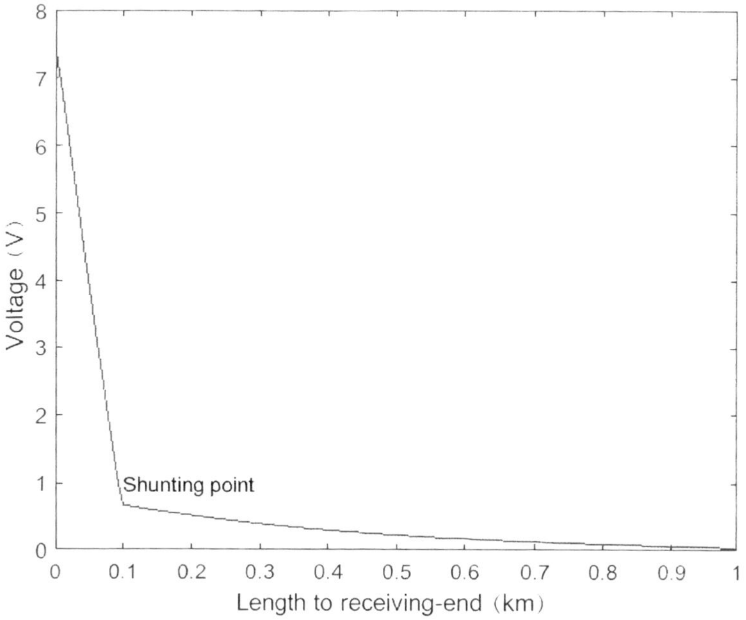

Figure 16.
Voltage distribution along the track circuit under shunted state.

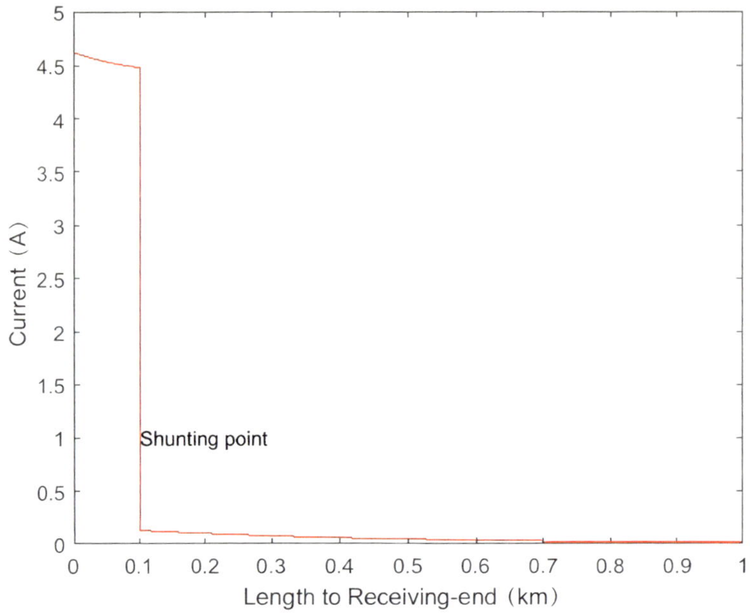

Figure 17.
Current distribution along the track circuit under shunted state.

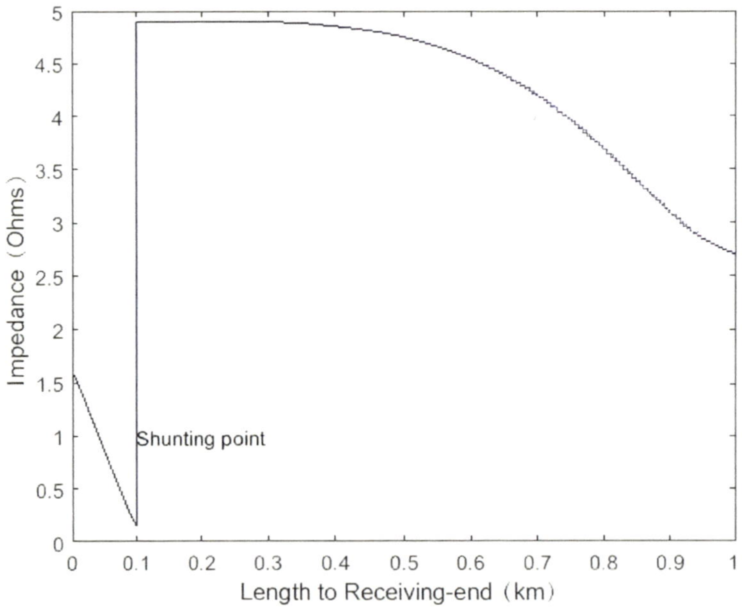

Figure 18.
Impedance distribution along the track circuit under shunted state.

6 Conclusion

This study provides a comprehensive investigation into the integration of track circuit technology within China's all-electronic interlocking framework, proposing a unified solution to address the historical fragmentation between traditional station-based systems (e.g., 25 Hz phase-sensitive) and section-based systems (e.g., ZPW-2000). The research outlines the limitations of electromechanical relays in conventional 25 Hz track circuits, including low return coefficients and susceptibility to 50 Hz interference, and demonstrates how microelectronic receivers (such as the JXW25) substantially improve reliability and anti-interference performance. A key innovation presented is the implementation of the ZPW-2000 track circuit within station areas, forming an integrated architecture that employs mechanical insulation joints and customized impedance-matching equipment to support continuous train–ground information transmission – a critical requirement for advanced train control systems.

Furthermore, the development of a domestically produced, SIL 4-compliant electronic control module with a 2oo2 architecture is described, representing a major advancement in replacing traditional relay-based interfaces and promoting the transition to fully electronic, intelligent interlocking systems. The study also examines mechanisms of adjacent-section and adjacent-line interference, emphasizing the importance of maintenance practices and understanding coupling effects to ensure operational safety. Through mathematical modeling and empirical simulations, the transmission characteristics are quantitatively analyzed, confirming that signal attenuation is influenced by carrier frequency, ballast resistance, and rail length. The circuit's performance is further validated by a significant voltage attenuation ratio under shunted conditions. In summary, this integrated technological approach establishes a robust, standardized, and highly reliable foundation for the future of digitalized railway signaling in China.

Acknowledgments

This study was sponsored by the Key Project of China Academy of Railway Sciences Co., Ltd. (No. 2024YJ193) and the Key Project of Beijing Huatie Information Technology Co., Ltd. (No. 2024HT10).

Conflict of interest

The authors declare no conflict of interest.

Author details

Haonan Feng[1]*, Yingjie Ji[2], Zhiguo Liang[1], Hongli Guo[1] and Xiaojiao Ma[3]

1 Signal and Communication Research Institute, China Academy of Railway Sciences Corporation Limited, Beijing, China

2 China Railway Chengdu Group Corporation Limited, Chengdu, China

3 China Railway Test & Certification Center Limited, Beijing, China

*Address all correspondence to: fhn02212005@163.com

References

[1] Haonan F, Kejia X, Sunjianguo SM. A domestic electric control module for 25Hz track circuit. Foreign Electronic Measurement Technology. 2021;**10**: 60–65. DOI: 10.19652/j.cnki.femt.2102863

[2] Zhiguo L.Metro deport's computer interlocking system adapting to electronic executive unit. Urban Rapid Rail Transit. 2016;**29**(04):59–63

[3] Haonan F. Design and research on all electric computer interlocking system for urban transit. Journal of Railway Science and Engineering. 2021;**08**:2145–2155. DOI: 10.19713/j.cnki.43-1423/u.T20200965

[4] Mingjun Y. Reserach on TYJL-IIIE all-electronic computer interlocking system and its application. Railway Signalling and Communication. 2022;**58**(11):29–37+41. DOI: 10.13879/j.issn.1000-7458.2022-11.22171

[5] Rui M, Chunfeng H, Xiaobo Y. Analysis and application of all-electronic frequency shift coding drive module. Railway Signalling and Communication. 2021;**57**(02):23–26+48. DOI: 10.13879/j.issn.1000-7458.2021-02.20388

[6] Haonan F, Yunhun G, Wei F, Ming P. Design and research of domestic all-electric signal control unit. Chinese Journal of Electron Devices. 2021;(05):1060–1065

[7] Haixu Z. Design of an all-electronic high-voltage pulse track circuit system. Railway Signalling and Communication. 2020;**56**(12):38–40. DOI: 10.13879/j.issn1000-7458.2020-12.20260

[8] Yunhun G, Mengyao Z, Anping H. Digital interlocking system based on intelligent trackside equipment. China Railway. 2020;(02):6–10. DOI: 10.19549/j.issn.1001-683x.2020.02.006

[9] Cuiqin Z. Research on Adjacent Crosstalk Based on ZPW-2000 Track Circuit Used in Stations and Intervals [thesis]. Lan Zhou: Lanzhou Jiaotong University; 2020

[10] Yan Y. The Modeling and Simulation of Track Circuit [thesis]. Cheng Du: University of Electronic Science and Technology of China; 2005

[11] Gang Z. Modeling and Model Verification of Uninsulated Track Circuit [thesis]. Nan Chang: East China Jiaotong University; 2022

[12] Guangzhi M, Xueshu X. The modeling and simulation of track circuit. Electric Drive for Locomotives. 2004;**01**:41–43+46. DOI: 10.13890/j.issn.1000-128x.2004.01.012

Section 3

Sustainable Mobility, Operations, and Logistics

Chapter 4

A Review of Sustainable Location–Allocation Models for City Logistics Using High-Speed Rail: Case Study of the Ottawa–Quebec Corridor and Greater Montreal Area

Yong Lin Ren and Anjali Awasthi

Abstract

This review examines sustainable location–allocation models for city logistics leveraging high-speed rail (HSR), with a case focus on the Ottawa–Quebec corridor and Greater Montreal Area (GMA). We address three research questions: How do metaheuristics and Graph Neural Network (GNN-RL) optimize HSR-based logistics hubs? What are the comparative strengths of p-median, hub-and-spoke, and maximal coverage models in mid-density corridors? How can insights from global HSR logistics networks inform North American planning? A systematic bibliometric analysis (2014–2024) of Scopus and Web of Science datasets, processed via VOSviewer, identifies trends in algorithmic applications (e.g., genetic algorithms, neural networks). Case studies from China, Japan, and the EU are contrasted with GMA's nascent HSR logistics potential. Findings highlight trade-offs between computational efficiency and solution robustness, offering policy recommendations for sustainable freight integration.

Keywords: HSR, city logistics, location–allocation, neural networks, GMA

1. Introduction

In the contemporary era of rapid urbanization and evolving transportation landscapes, high-speed rails (HSRs) have emerged as a transformative force in city logistics. This review centers on sustainable location–allocation modeling for city logistics facilitated by HSR, a multifaceted area of study that encompasses three core subproblems: the location problem, the allocation problem, and the mathematical modeling problem. The exponential growth of the HSR system in recent years has redefined the boundaries of traditional transport and logistics. It has not only

IntechOpen

revolutionized the way people travel but also reshaped multimodal transportation strategies, opening up new avenues for efficient goods transportation within and between cities. This transformation has captivated the attention of numerous researchers, leading to a surge of studies in the field of HSR-based city logistics location–allocation.

The location problem lies at the heart of optimizing HSR-based city logistics networks. It involves determining the most suitable sites for hub centers, which act as crucial nodes in the distribution process. These hubs are not just simple transfer points; their location has far-reaching implications. It affects the volume of goods that can be efficiently distributed, the time taken for auxiliary transportation to reach end users, and the overall loading capacity of the HSR. For example, a well-placed hub can reduce delivery times, lower transportation costs, and enhance the overall efficiency of the logistics system. Conversely, a poorly located hub may lead to bottlenecks, increased costs, and inefficiencies in the supply chain.

The allocation problem, on the other hand, is concerned with the optimal distribution of goods from the hub centers to their final destinations, whether they are individual customers or groups. This requires making informed decisions about which goods to allocate, in what quantities, and through which routes. Factors such as delivery time, cost, customer demand, and the capacity of the transportation network all come into play. An effective allocation strategy can ensure that goods reach their destinations promptly while also minimizing costs and maximizing customer satisfaction.

Both the location and allocation problems necessitate the use of mathematical modeling. Mathematical models serve as powerful tools to represent the complexity of the location–allocation problem accurately. They construct objective functions that can generate optimal or near-optimal solutions. In this review, we will explore a variety of metaheuristic algorithms proposed by researchers. Metaheuristics, such as the genetic algorithm (GA), ant colony optimization (ACO), and Tabu Search, have shown great promise in solving complex optimization problems in logistics. GA, inspired by the principles of natural evolution, can quickly search for optimal solutions under suitable conditions. ACO, inspired by the behavior of ants, is highly effective in vehicle routing problems within advanced location–allocation systems. Tabu Search, with its ability to avoid revisiting unproductive solutions, helps in finding better solutions in a shorter time.

In addition to metaheuristics, we will also delve into the role of artificial intelligence (AI) and machine learning techniques. Linear regression (LR), neural algorithm (NA), TreeNet algorithms (TAs), and neural networks (NNs) are among the AI methods that have been applied to address the location–allocation problems in HSR-based city logistics. These AI techniques can analyze large volumes of data, identify patterns, and make accurate predictions, enabling more intelligent decision-making in logistics operations. Evaluating the sustainable location–allocation for city logistics using HSR is also a critical aspect of this review. We will discuss various evaluation approaches, including multi-objective optimization, partial derivative methods, and potentially mixed methods. These evaluation methods help in assessing the effectiveness, efficiency, and sustainability of different location–allocation strategies.

A systematic literature review was conducted using Scopus and Web of Science to identify trends in HSR-based city logistics, with a focus on metaheuristics and AI applications. By studying the research findings of previous scholars, we aim to synthesize the existing knowledge, identify research gaps, and provide valuable insights into the future direction of sustainable location–allocation for HSR-based city logistics. Additionally, this review will specifically focus on the Greater Montreal Area (GMA), exploring the unique challenges and opportunities it presents in the context of HSR-based city logistics location–allocation. While research in this region remains limited compared to countries such as China and Japan, there is growing interest in leveraging HSR infrastructure to support sustainable freight movement in mid-sized metropolitan areas.

The Ottawa–Quebec City corridor offers a strategic setting for studying HSR-based logistics due to its dense urban centers, intermediate cities, and existing intercity rail infrastructure. Insights drawn from this case study will be framed as transferable best practices applicable to other North American corridors facing similar logistical and environmental challenges. By integrating data-driven forecasting, algorithmic optimization, and policy analysis, this review contributes to advancing the understanding of how HSR can be effectively integrated into sustainable city logistics planning. It aims to inform both academic discourse and the practical implementation of HSR-based freight systems in emerging urban contexts.

In the era of rapid urbanization and climate-conscious transportation planning, HSR has emerged as a transformative solution for sustainable city logistics. This review investigates the application of location–allocation models to optimize HSR-based freight networks, with a focus on the Ottawa–Quebec corridor and the GMA. These regions were selected due to their (a) intermediate city density, (b) existing rail infrastructure, and (c) alignment with Canada's 2030 HSR feasibility goals. Unlike high-density Asian corridors (e.g., Beijing–Shanghai or Tokyo–Osaka), this region exemplifies the challenges of scaling HSR logistics for mid-sized markets, where freight demand is dispersed and infrastructure integration requires careful planning.

1.1 The role of HSR in city logistics

HSR systems, traditionally designed for passenger transport, are increasingly being repurposed for freight to reduce road congestion and carbon emissions. In Europe and Asia, dedicated HSR freight services (e.g., France's TGV Fret and China's CRH Express) demonstrate the viability of rail-based urban logistics. However, North America lags behind, with limited studies on HSR's potential for freight in mid-density corridors. The Ottawa–Quebec corridor – an 800 km stretch connecting major economic hubs – presents a unique opportunity to address this gap. With Canada's federal government evaluating HSR feasibility under the 2030 Transportation Green Plan, optimizing logistics networks alongside passenger services is critical.

Sustainable location–allocation for HSR-based city logistics involves three interdependent challenges:

1. Location problem (LP): Identifying optimal hub locations (e.g., intermodal terminals) to minimize transit times and maximize coverage.

2. Allocation problem (AP): Distributing goods efficiently from hubs to last-mile destinations under capacity constraints.

3. Modeling problem: Selecting algorithms that balance computational efficiency with real-world applicability.

To address these, we formulate three research questions:

- RQ1: How do metaheuristics (e.g., GAs, ACO) compare in solving HSR hub location problems (HLPs)?
- RQ2: What allocation strategies (e.g., hub-and-spoke, p-median) best suit mid-density corridors like the GMA?
- RQ3: Can AI-driven models (e.g., NNs) enhance decision-making for stochastic freight demand?

1.2 Methodological approach

This review combines the following:

- Bibliometric analysis: A systematic review of 2,639 Scopus/WoS publications (2014–2024) using VOSviewer to map keyword cooccurrence (e.g., "HSR," "location allocation," "city logistics").
- Case study synthesis: Lessons from China (over 60,000 km HSR network) and Japan (dedicated overnight freight Shinkansen) are contrasted with the GMA's nascent projects, such as the proposed Toronto–Quebec HSR.
- Algorithm benchmarking: Computational trade-offs (e.g., solution quality vs. runtime) are evaluated for p-median, maximal coverage, and stochastic models.

Why the Ottawa–Quebec corridor and GMA?
The selection of these regions is strategic:

1. Intermediate density: Cities such as Ottawa, Montreal, and Quebec have populations of 1–4 million – sufficient for freight demand but lacking the scale of megacities. This tests HSR's viability in "second-tier" markets.
2. Existing infrastructure: Via rail's corridor services and Montreal's intermodal hubs (e.g., Dorval Logistics Park) provide a foundation for integration.
3. Policy alignment: Quebec's 2030 Sustainable Mobility Plan prioritizes rail decarbonization, while the Boston-Montreal HSR (BMHSR) Feasibility Study (Parsons and Douglas, 2003) identifies freight synergies.

Unlike China's centralized planning, North American logistics face fragmented governance and reliance on trucks. For example, Montreal's port handles 2.5 million TEUs annually, yet 70% of freight moves by road. HSR could alleviate this if hubs are optimally placed – a challenge this review addresses, as the distribution of those cities as shown in **Figure 1** along the Ottawa–Quebec City corridor highlights. To make intermodal logistics using HSR viable, its urban–suburban specialty must be

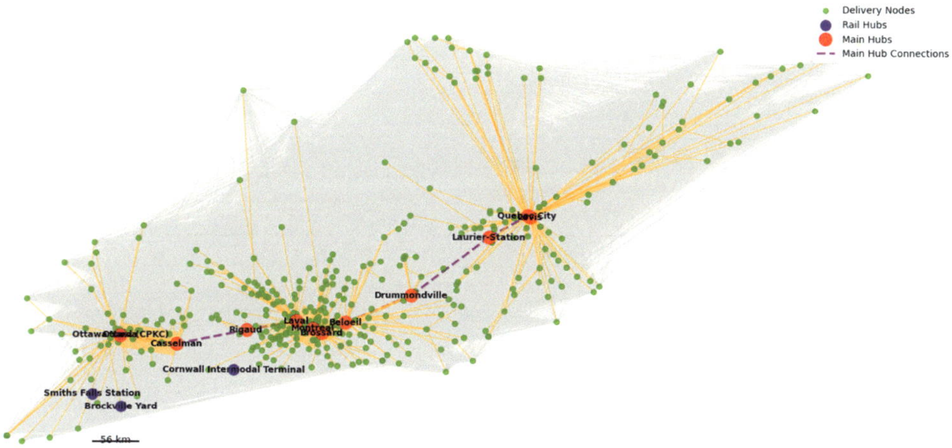

Figure 1.
Cities in the Ottawa–Quebec City corridor.

considered, along with factors such as winter weather, wetlands, and protected areas.

1.3 Contribution to literature

Prior studies focus on high-density Asian/European networks, neglecting mid-density adaptations. By analyzing the Ottawa–Quebec corridor, we:

- Propose transferable models for regions with dispersed demand;
- Highlight policy barriers (e.g., shared passenger-freight tracks); and
- Benchmark algorithm performance using real-world constraints (e.g., winter weather disruptions).

1.4 Structure of the review

Section 2 details the methodology, including bibliometric protocols. Section 3 evaluates metaheuristics and AI for location–allocation. Section 4 synthesizes global case studies, while Section 5 focuses on GMA-specific insights. The conclusion outlines recommendations for practitioners. This review aims to bridge the gap between theoretical models and practical HSR logistics deployment, offering a roadmap for sustainable freight in mid-sized corridors.

2. Research methodology and study conducted on sustainable location–allocation problems for city logistics using HSR

This systematic review was conducted following the PRISMA (shown in **Figure 2**) guidelines to ensure a comprehensive and transparent literature selection process. The initial search across the Scopus and Web of Science databases yielded

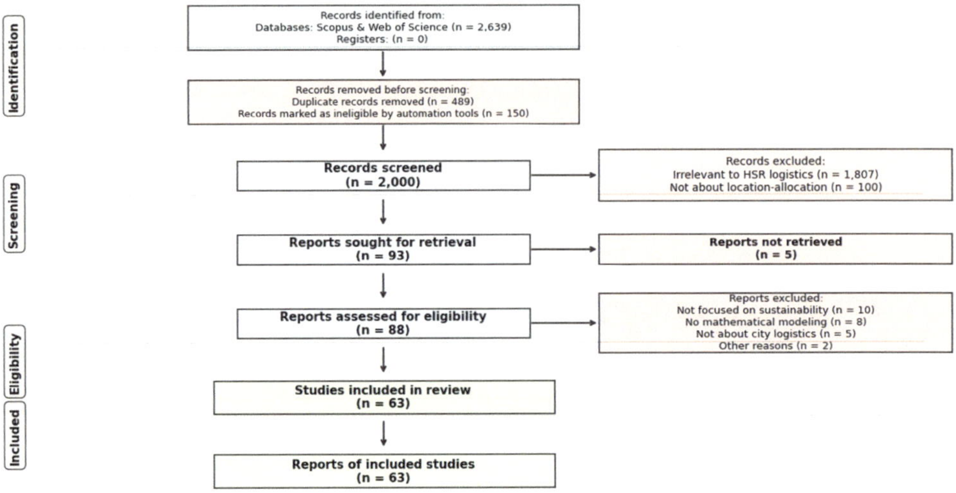

Figure 2.
PRISMA flowchart of the review.

2,639 records. These records underwent a rigorous multi-stage screening and eligibility assessment to identify studies relevant to sustainable location–allocation models for HSR in city logistics. After the removal of duplicates, screening of titles and abstracts, and a full-text review based on strict inclusion criteria, a final corpus of 63 studies was selected for in-depth analysis within the review. This refined selection enabled a focused examination of algorithmic applications, case studies from global HSR networks, and their specific implications for the Ottawa–Quebec corridor and GMA.

Of the 2,639 publications initially identified from the Scopus and Web of Science databases, the geographic distribution of research output, as visualized in **Figure 3**, reveals a pronounced global disparity. This analysis shows a dominant concentration of relevant studies originating from Asia, particularly China, a finding that aligns directly with the nation's extensive investment and deployment of HSR infrastructure. The significant correlation between a country's HSR network length and its volume of published research underscores the role of real-world implementation in driving academic inquiry. This geographical context is crucial for framing the subsequent bibliometric analysis, as it highlights the regions from which the predominant methodologies and case studies – essential for informing the Ottawa–Quebec and GMA case – are derived.

This review on sustainable location–allocation problems in city logistics using HSR encompasses the location problem, the allocation problem, and the modeling aspect for HSR-based city logistics location–allocation. To comprehensively understand these, a multi-faceted research methodology is employed, which includes identifying relevant metaheuristic algorithms and AI applications, as well as conducting bibliometric and visualization analyses of publications and studies related to HSR-based city logistics.

Metaheuristics play a crucial role in solving the location–allocation problem for sustainable and advanced city logistics. Many researchers have proposed various metaheuristic algorithms. For instance, Gogna et al. [1] and Pan [2] have contributed

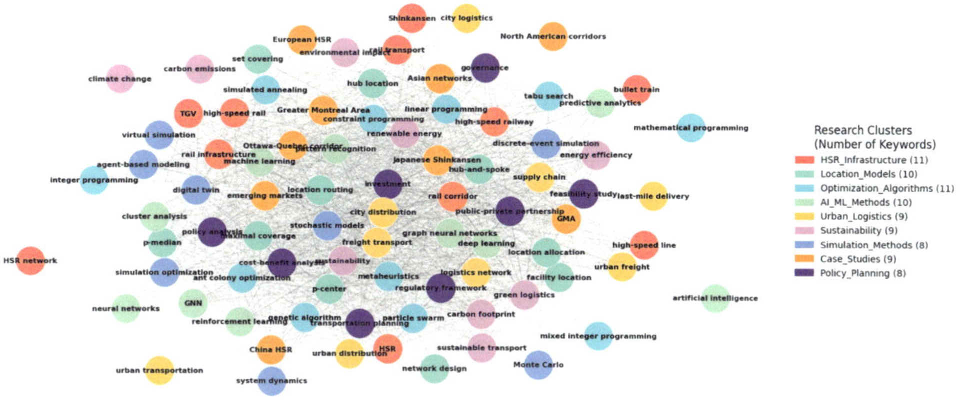

Figure 3.
Analysis of 2,639 publications from Scopus and Web of Science.

to this field. These algorithms often draw on AI-related concepts such as bio-inspired computations and swarm intelligence, as described by Bonabeau et al. [3] and Good and Brown [4]. Examples of such algorithms include the glowworm algorithm by Krishnanand and Ghose [5], bacteria foraging, and particle swarm optimization by Shi and Eberhart [6]. Additionally, Pareto-based algorithms proposed by Corne et al. [7, 8] are also relevant.

In terms of allocation methods, metaheuristics have been widely used. Hruschka et al. [9] and Biswas et al. [10] categorized algorithms into overlapping, hierarchical, and partitional types. Karaboga [11, 12] introduced harmony search-based partitional algorithms. The GA, inspired by evolutionary theory [13], has been applied in various supply chain contexts. For example, Reeves [14] and others applied the integer encoding of chromosomes in supply chain problems. GA can perform quick calculations under suitable conditions for specific objectives. The simulated annealing (SA) algorithm, a stochastic global optimization method proposed by Ali and Storey [15], is also used. Tabu Search (TS), developed by Glover [16, 17], is designed to avoid revisiting unproductive solutions during the optimization process. ACO has proven efficient for vehicle routing in advanced location–allocation systems [18–20]. It has also been studied for its ability to optimize supply chain management and solve the traveling salesman problem (TSP), as demonstrated by Gambardella et al. [21].

These metaheuristics provide mathematical frameworks for advanced location–allocation in city logistics. For example, Kansou and Kanso [22] developed a mathematical model integrated with a hybrid metaheuristic to solve the periodic capacitated arc routing problem (PCARP). Zhang et al. [23] further enhanced this strategy by proposing new route decomposition and a memetic algorithm (MA) framework for PCARP. NNs with probabilistic algorithms, such as those by Specht [24], have also been explored in this context. Metaheuristics for location–allocation in city logistics are increasingly being developed for AI-enabled operational execution in delivery processes.

In the shipping logistics field, service quality measurement is vital for analyzing advanced location–allocation. The highly competitive delivery industry requires strategies to increase profits for both customers and suppliers. Good and Brown [4] proposed a multi-objective design for an advanced location–allocation delivery ship, considering capabilities, design parameters, military effectiveness, costs, and risks.

Nam and Song [25] investigated Korean shipping location–allocation, which is influenced by downstream demands in international trade. Miyashita [26] studied structural changes in international advanced location–allocation and found that mid- to long-term contracts can enhance shipper satisfaction by ensuring long-term profit stability. In air logistics, Murakami [27] in Japan researched market performance related to low-cost entries into the airline industry. They analyzed cargo growth trends and discussed the economic implications of large containerships on port operations and container trade patterns. Chung [28] studied strategic business types within the air-location allocation industry, classifying related industries into air-location allocation-dependent and air-location allocation-relevant sectors. Purkayastha et al. [29] focused on the collection and recycle bin location–allocation problem in solid waste disposal. Their research aimed to optimize the location–allocation of collection and recycling bins to reduce the number of dumping yards and promote sustainable development. They considered factors such as vehicle location, minimizing total investment cost, dissatisfaction, and the number of collections, while maximizing the population served by recycling drop-off stations. They concluded that there is a lack of research on this problem, and given the significant budget for waste collection, more attention should be paid to it for environmental protection and cost reduction.

Sustainable advanced location–allocation for city logistics has been extensively researched by many scholars. A key question is how to minimize the time required to provide optimal location–allocation services. For example, Meituan employs smart technologies to achieve an average delivery time of 10–30 minutes from providers to customers. This system utilizes AI to coordinate the various stakeholders – including managers, delivery personnel, providers, and customers – benefiting all participants in the network. However, sustainable location–allocation in suburban areas poses different challenges compared to urban environments, often due to the sparsity and complexity of these regions. To address these challenges, Japan constructed a HSR between Tokyo and Osaka in 1959, significantly improving location–allocation in suburban areas. Today, from Shanghai to Beijing, covering 1,213 kilometers, the journey takes just 5 hours to complete the whole travel, including the station stop time. Currently, the China High-Speed Railway (CHSR) system, with 45,000 kilometers, not only enhances the location–allocation efficiency of logistics in remote city areas but also offers sustainable development solutions for city logistics and rural communities. For the GMA region, Parsons and Douglas [30] suggested a report about the BMHSR planning that provided a detailed description of city logistics using HSR.

Lyu et al. [31] aimed to develop a routing architecture integrating a discrete-event simulation (DES) last-mile delivery model. They used a stochastic two-tier hub-and-spoke (H&S) architecture approach and DES to optimize urban freight logistics. DES is a valuable technique for modeling the complex and dynamic nature of logistics operations with stochastic elements. By considering factors such as hub centers, truck allocation, and goods delivery, DES can evaluate different scenarios and their impacts on key performance indicators such as delivery cost, resource quantity, and time efficiency. It also provides a robust framework for clarifying operational processes. The two-tier H&S architecture integrated with a DES model helps address the stochastic nature of freight delivery. Their calculations determined parameters for cluster determination, with the initial iteration (*N*min = 20, Research = 2000) resulting in four clusters, one of which was dislocated and retained as an independent cluster. The remaining three clusters were then reanalyzed in Iteration 2 (*N*min = 250, Research = 1250).

Many scholars have extensively studied sustainable advanced location–allocation for city logistics. For example, Meituan uses smart technologies to achieve an average delivery time of 10–30 minutes, coordinating various stakeholders through AI. However, sustainable location–allocation in suburban areas presents different challenges due to their sparsity and complexity. Japan's 1959 construction of an HSR between Tokyo and Osaka improved location–allocation in suburban areas. CHSR system, with 45,000 kilometers, not only enhances logistics location–allocation efficiency in remote areas but also offers sustainable development solutions for city logistics and rural communities. In the GMA, Parsons and Douglas [30] reported on the BMHSR planning, which described city logistics using HSR. Zhang et al. [32] studied the HSR system's impact on the improvement of factor allocation efficiency between regions and urban environmental quality. They used the propensity score matching (PSM)–DID method to eliminate the selective bias problem of HSR openings and reselected control group samples for comparison. They found that the average misallocation of capital factors was 18.69%, and labor misallocation was the main cause of factor misallocation between prefecture-level cities in China. However, research on the sustainable location–allocation problem for city logistics using HSR, especially in the context of metaheuristics and AI-based optimization approaches, is still limited. Only a few studies focus on using AI algorithms to optimize city logistics using HSR. City logistics also deals with issues such as traffic congestion and vehicle circulation capacity. Khosravi et al. [33] studied the hub-and-spoke network design considering congestion and flow-based cost functions. They applied linearization techniques to linearize the model and proposed tightening rules to refine the model and assist in the solution process. Lyu et al. [31] proposed a stochastic two-tier H&S framework for urban freight optimization, leveraging DES to model operational variability and density-based spatial clustering to identify efficient delivery zones. Their findings confirmed that integrating stochastic modeling, spatial cluster analysis, and simulation techniques enables logistics practitioners to optimize truck allocation, reduce empty mileage, and establish adaptive delivery networks in new urban or suburban areas. The study emphasized that this integrated approach improves resource utilization by 14–22% compared to traditional point-to-point routing, particularly in regions with uneven demand distribution. Liu et al. [34] studied the city logistics of Nanjing, China, through the design of an underground logistics network for large-scale municipal solid waste collection. Their study proposed an automated, large-scale underground collection system for MSW, offering a new perspective on sustainable city development and underground space utilization, and enabling reasonable location–allocation for municipal solid waste delivery.

Table 1 shows synthesized algorithmic applications in urban logistics across metaheuristics, NNs, and HSR integration, offering fertile ground for deeper research. Metaheuristics such as ACO address NP-hard problems (e.g., TSP, location–allocation), while NNs – from self-organizing feature maps with guided refining (SOFMGR) procedure to radial basis function neural networks (RBFNNs) – enable data-driven solutions for complex spatial tasks, including emergency evacuation and wind farm allocation. HSR logistics studies, such as the HLP model

Algorithms/ city logistics	Authors/ coauthors	Solved problem	Methodology/solution
Metaheuristics	Gogna and Tayal [1]	Location–allocation	Applies metaheuristics for location–allocation.
	Marco [19]	Optimization	Ant colony optimization/ACO.
	Subramanyam and Gounaris [18]	TSP	TSP (hard problem).
NNs /AI	Karaboga and Basturk [12]	XOR problem	NN.
	Mahfuzur et al. [35]	Emergency evacuation	Location–allocation GIS sensing.
	Tsadiras et al. [36]	Buffer allocation problem	NN-based decision support system.
	Hsieh and Fang-Chih Tien [37]	Location–allocation problems with rectilinear distances	A SOFM artificial neural network with a new guided refining [SOFMGR] procedure.
	Xenitidis et al. [38]	Wind farms allocation	Using radial basis function neural networks [RBFNNs].
	Zhang et al. [39]	Resource allocation	A recurrent neural network-based trajectory prediction method.
City logistics with HSR	Li et al. [40]	HSR location	HSR HLP.
	Zhen et al. [41]	Delivery problem	City logistics using HSR.
	Parsons and Douglas [30]	HSR transportation	BMHSR

Table 1.
Algorithm for location–allocation in city logistics.

and Boston-Montreal corridor analysis, reveal the infrastructure's role in reshaping delivery networks. More inquiry could explore hybrid metaheuristic-AI frameworks for dynamic scheduling or develop explainable AI for neural logistics models, bridging algorithmic rigor with real-world transport systems.

In recent years, numerous researchers have explored the location and allocation problems in city logistics using HSR through various models related to HLP, hub location–allocation problem (HLAP), and location–allocation with HSR (LAHSR) problems. This area of study has become a significant focus in the logistics field. For instance, in the context of the HLP, Pašagić et al. [42] considered the location of urban logistic terminals as a HLP. They proposed that such location problems can be solved in practice using heuristic methods or a combination of heuristic and exact methods. Shahparvari et al. [43] adopted a GIS-LP integrated approach and used mixed-integer quadratic constrained programming (MIQCP) and a K-means heuristic to solve the problem. Wang et al. [44] combined the MA and Monte Carlo (MC) simulation for their proposed stochastic model, and based on the Civil Aeronautics Board (CAB) and Turkish network datasets, they obtained computational results. Anderluh et al. [45] utilized the analytic hierarchy process for hub location selection in a mid-hub (a medium-sized city hub).

Regarding the HLAP, Ishfaq and Sox developed a mathematical model using the multiple-allocation p-hub median. They applied this model to large-size (100-node) problems and compared the solutions obtained through this meta-heuristic with tight lower bounds developed using a Lagrangian relaxation approach. Feng et al. [46] formulated a mixed-integer linear programming mathematical model and used the global optimization software Lingo. The optimization results of a real case in Xi'an verified the validity of their model. Li et al. [47] proposed an uncertain location–allocation model, taking into account three objectives: the urgency of emergency response, distribution costs, and carbon dioxide emissions. They found that the optimal solution to the models is sensitive to (g, w), and through numerical experiments, they generated different Pareto-optimal solutions by varying the values of g and w. Chu et al. [48] combined the firefly algorithm and GA to create a novel and effective IFA–GA algorithm with three schemes. They tested it on a small-scale instance with one SNS, 10 RTCs, and 40 LDPs, and the proposed model had 820 $(2 * 10 * 40 + 10 + 10)$ decision variables.

In the area of LAHSR, Ortega et al. [49] designed spatial data allocation procedures. They assessed the sensitivity of accessibility levels to both population distribution and geographical representation procedures by measuring access to the HSR network in Spain. Li et al. [50] used a mixed-integer programming node-arc model in addition to the macro-scale route. They employed the optimization solver Gurobi to solve the model and obtained the optimization scheme, taking the Zhengzhou railway hub as a case study. Repolho et al. [51] focused on two key strategic decisions for a given HSR line: station location and fleet composition. They approximated the interrelationships between strategic and subsequent tactical decisions and applied their method to the corridor for three fleet cost scenarios through the XPRES software and branch-and-cut techniques to determine optimal solutions. Loo [57] and Zhiran [52] conducted statistics on 1,627 HSR stations built from 1976 to 2021 in 533 Chinese and European cities. They identified node or activity clusters in cities through the street network. Their research indicated that determining the location of HSR stations is a complex issue. While they considered a series of factors related to city characteristics, HSR development, and urban form, other factors such as terrain, HSR services, economic functions, and planning regimes remained unexamined due to their complexity. Scopus bibliometric analysis was also carried out. From the statistics of keywords "location" + "allocation" + "HSR" in Scopus.com during 2014–2024 by top journals, data analysis for the location–allocation in city logistics using HSR was obtained.

Table 2 encapsulates research on logistics location–allocation problems, spanning HLP, HLAP, and LAHSR frameworks. In HLP, studies exhibit methodological sophistication: Pašagić et al. [42] operationalize heuristic-exact hybrid methods for urban terminal location, while Khosravi et al. [33] integrate GIS with MIQCP and K-means heuristics, addressing spatial-temporal complexity. Wang and Lo [53] advance stochastic modeling via MC simulation, and Anderluh et al. [45] deploy AHP for midi-hub strategic selection, highlighting multi-criteria decision-making. HLAP research demonstrates algorithmic innovation: Huang and Wu [54] addressed the challenge of validating meta-heuristic solutions for large-scale transportation networks by leveraging Lagrangian, while Feng et al. [46] apply mixed-integer linear programming to real-world logistics. Li et al. [50] pioneer triple-objective modeling (urgency, cost, sustainability) in emergency logistics, and Chu and Chen [55] developed an improved firefly algorithm-genetic algorithm (IFA-GA) hybrid algorithm, addressing decision variable complexity (820 variables in small-scale instances). LAHSR studies bridge infrastructure and logistics: Ortega et al. [49]

Problem	Authors/ Coauthors	Logistics model	Solution
HLP	Pašagić et al. [42]	Location of urban logistics terminals as a HLP.	The location problems of city terminals are hub problems that can be solved in practice by heuristic methods or a combination of heuristic and exact methods.
	Khosravi et al. [33]	GIS-LP integrated.	MIQCP and a K-means heuristic for solution.
	Wang and Lo [53]	Combining the MA and MC simulation for the proposed stochastic model.	Based on the CAB and Turkish network datasets for computational results.
	Anderluh et al. [45]	Analytic hierarchy location selection: a midi-hub – a medium-sized city hub.	Based on the analytic hierarchy process or the hub location selection.
HLAP	Ishfaq and Sox [56]	A mathematical model using the multiple-allocation p-hub median.	Large-size (100-node) problems. The solutions obtained using this meta-heuristic are compared with tight lower bounds.
	Feng et al. [46].	Mixed-integer linear programming mathematical model.	Global optimization software Lingo and the optimization results of a real case.
	Li. et al. [50]	Uncertain location–allocation model; three objectives are considered – (i) urgency, (ii) cost, and (iii) sustainability.	The optimal solution to models is sensitive to (g, w). In the numerical experiments, we will generate different Pareto-optimal solutions.
	Chu and Chen [55]	A novel, effective IFA-GA by combining the firefly algorithm and GA with three schemes.	A small-scale instance with one SNS, 10 RTCs, and 40 LDPs is used, and the proposed model has 820 $(2 * 10 * 40 + 10 + 10)$ variables.
LAHSR	Ortega et al. [49]	Design of spatial data allocation procedures.	Assess the sensitivity of accessibility levels to both population distribution and geographical representation procedures.
	Liu et al. [34]	Mixed-integer programming node-arc model, in addition to the macro-scale route.	Optimization solver Gurobi to solve the model with a case at Zhengzhou railway hub.
	Repolho et al. [51]	A given line focuses on two key strategic decisions: station location and fleet composition, with an approximate representation of the interrelationships between strategic and subsequent tactical decisions.	Applied to corridor COR8 for the three fleet cost scenarios specified earlier through the XPRES software and branch-and-cut techniques to determine optimal solutions.
	Loo and Zhiran [52]	Statistics with 1,627 HSR stations built from 1976 to 2021 in 533 Chinese and European cities. Nodes or activity clusters in cities are identified through the street network.	Putting the HSR station is a complex question, providing more empirical evidence across a wide range of Chinese and European cities over the last four decades.

Table 2.
Hub location–allocation and city HSR.

quantify HSR accessibility sensitivity, Liu et al. [34] optimize node-arc models for hubs, and Repolho et al. [51] integrate fleet composition with station location via XPRES software. Loo and Zhiran [52] provide cross-continental empirical evidence,

Algorithm class	Strengths in HSR logistics	Weaknesses in HSR logistics
Metaheuristics (GA, ACO) Global optimization for NP-hard problems (e.g., HSR hub location with nonlinear constraints). Interpretable decision processes (e.g., ACO's pathfinding mimics real-world freight routing).	High computational cost for real-time adjustments (e.g., GA requires 100 + iterations for large networks). Sensitive to parameter tuning (e.g., mutation rate in GA affects solution diversity).	Strategic planning: Long-term HSR hub location–allocation [40, 47].
Neural networks (RNN, GNN) Data-driven prediction for stochastic freight demand (e.g., RNNs model time-series traffic patterns). Parallel processing for real-time routing (e.g., GNNs optimize node-arc networks in HSR corridors).	Black box nature hinders policy interpretation (e.g., it is difficult to explain why a hub location is chosen). Require large labeled datasets (e.g., HSR freight flow data, less so in North America).	Tactical operations: Dynamic allocation with HSR-feeder integration (Zhen et al. 2019) [41].
Hybrid models (Metaheuristic + AI) Combines global optimization with data adaptivity (e.g., ACO-GNN hybrid for HSR station location under uncertain demand).	Complex implementation (e.g., synchronizing heuristic search with NN predictions).	Adaptive systems: HSR logistics networks require both optimality and real-time learning [55].

Table 3.
Algorithm comparison for the location–allocation of the HSR system.

though terrain and planning regimes remain underexplored. Collectively, these works underscore the need for hybrid algorithmic frameworks, multi-objective optimization, and big data integration in next-generation logistics research.

Table 3 presents a comparison of algorithms for the location–allocation of the HSR system. The HSR-specific models (HLP, stochastic hub-spoke) strengthen HLP models, as seen in Li et al. [40, 47], aligning with China's corridor-specific infrastructure and policy alignment. Similarly, the stochastic models in Lyu et al. [31] reflect real-world constraints (e.g., Montreal's winter road closures). Additionally, these models mitigate regional bias, as models developed for Asia may not translate well to North America's fragmented governance, as noted by Parsons and Douglas [30]. Furthermore, many HLP models overlook dynamic factors such as HSR service disruptions (e.g., snowstorms in the GMA).

3. Scopus bibliometric analysis

Scopus bibliometric analysis inclusion criteria are as follows:

- Keywords (Boolean search string): They are ("location allocation" OR "hub location" OR "logistics network design"), ("high-speed rail" OR "HSR" OR "TGV" OR "Shinkansen"), and ("Montreal" OR "Ottawa–Quebec corridor" OR "GMA").
- Document types: Peer-reviewed journal articles, conference proceedings, and book chapters (excluding editorials, letters, and technical reports).

- Language: English-language publications (due to the dominant academic discourse in logistics research, but non-English studies on Asian HSR systems were manually reviewed for relevance).

- Time range: 2014–2024 (to capture post-2014 HSR expansion in China and the EU, and align with GMA's emerging HSR initiatives).

To further explore the research landscape, a search was conducted on Webofscience.com using the keywords "location or allocation" + "city logistics" + "HSR" + "Montreal" for the period 2014–2024. By selecting "highly cited papers" + "hot papers" and filtering with relevant subject areas such as "environmental science," "engineering environmental," "computer science artificial intelligence," "mathematics applied," "computer science information systems," "mathematics," "transportation," "sustainable cities and communities," and "affordable and clean energy," 2,639 publications were retrieved from the Web of Science Core Collection.

Figure 4 shows the distribution of the number of articles by country. The Asian region, especially China, has published the most relevant research results, indicating its advanced position in HSR-based city logistics location–allocation. The number of published papers shows that the more kilometers of HSR, the more papers are published. The United States also has significant research in this domain due to its high-tech capabilities, despite not having an extensive HSR system. Europe and Japan have excellent research based on their advanced HSR systems, and Saudi Arabia has also contributed with its HSR-related research.

Figure 5 shows the distribution of research in different categories. The main research areas are computer science, AI, mathematics combined with other sciences, transportation science, engineering, and environmental science, which implies that high-tech is more widely applied in the development of the HSR system.

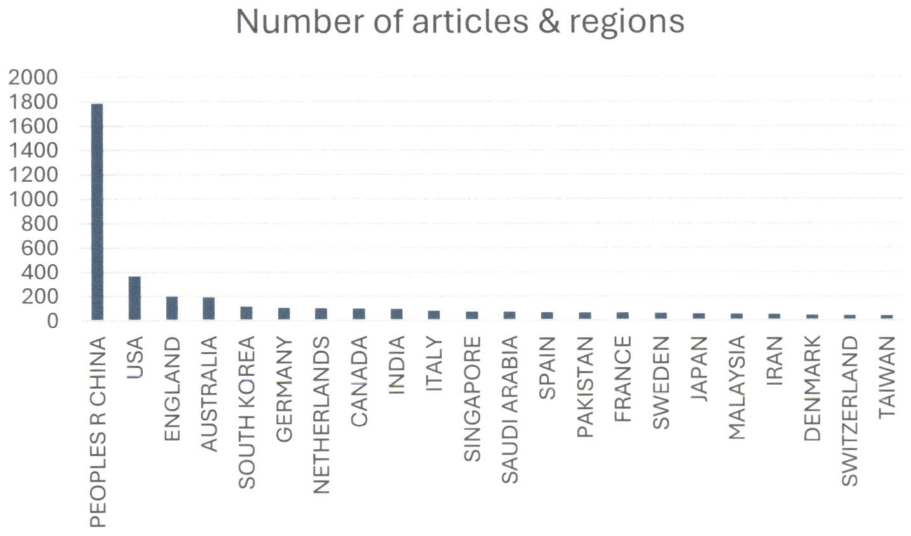

Figure 4.
Number of articles in different regions by limited keywords.

Figure 5.
Research category distribution: Top fields include Computer Science (Artificial Intelligence), Mathematics, Transportation Science, and Environmental Science.

In addition, a Scopus bibliometric analysis was carried out. Using the keywords "location" + "allocation" + "HSR" in Scopus.com for the period 2014–2024, data analysis was obtained. **Figure 6** shows the research on location–allocation for city logistics using HSR by institutes. The top 8 researchers and authors are from Chinese universities, indicating their dominance in this research field and their advancements in many areas of science and technology, with Beijing Jiaotong University being ranked No. 1, focusing on the problem using HSR.

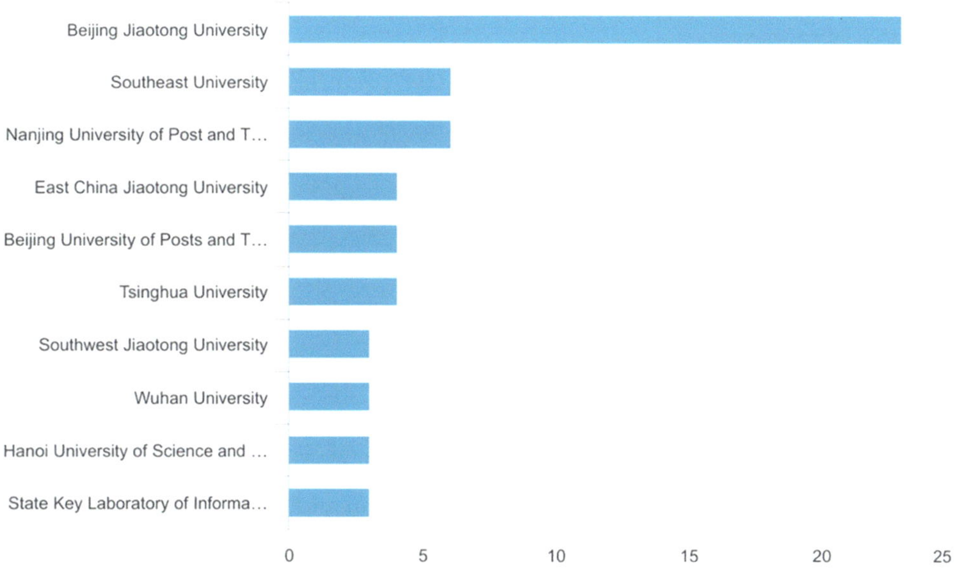

Figure 6.
Location–allocation problem using city HSR vs. institute.

Figure 7 shows the distribution of research in various countries, which are more active and advanced in HSR-based city logistics research, particularly in China. It also shows that there is some research in the United States, but there is not much actual activity there.

However, many other countries, such as Vietnam, Canada, Australia, and others, have conducted research on the challenges of using HSR, which demonstrates their interest in developing an HSR system.

Figure 8 shows that the majority of research in this area, when using the keywords "location" or "allocation" + "HSR" from 2014–2024 in famous journals, falls within the computer science, mathematics, and engineering fields. More than 67.5% of the subjects belong to these areas. The analysis also shows that the location–allocation problem using the city HSR system mainly applies various algorithms for solutions. The analysis further reveals that studies addressing location–allocation problems in urban HSR systems primarily employ various algorithms for problem-solving. Additionally, focusing on urban logistics integrated with HSR systems, a search using the keywords "city logistics" + "HSR" generated **Figure 9**, which indicates that 3,913 publications have been contributed by researchers. These data underscore a growing trend of increasing scholarly attention toward urban logistics applications of HSR systems. **Figure 10** shows the papers published in the last 10 years, indicating that the research is becoming increasingly attractive to many people.

When using the keywords "location" or "allocation" + "HSR" + "logistics," **Figure 10** shows the number of publications vs. the year 2014–2024 by the top 20. It demonstrates that many researchers are interested in the logistics of using the HSR system in this field.

For the location–allocation for city logistics using HSR in the GMA, there has been limited research. However, in the past five years, with the use of keywords

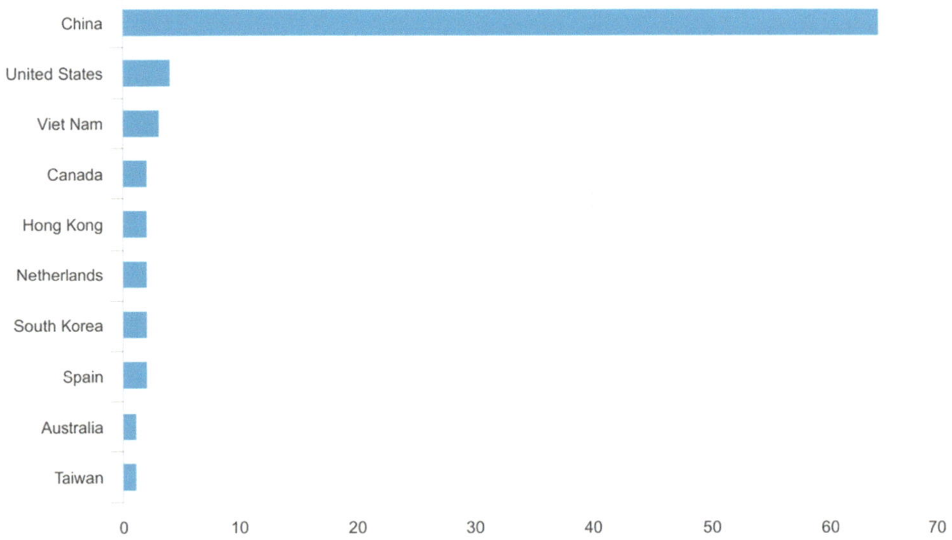

Figure 7.
National research distribution for HSR city logistics: China has the highest output, followed by the USA, Vietnam, and Canada.

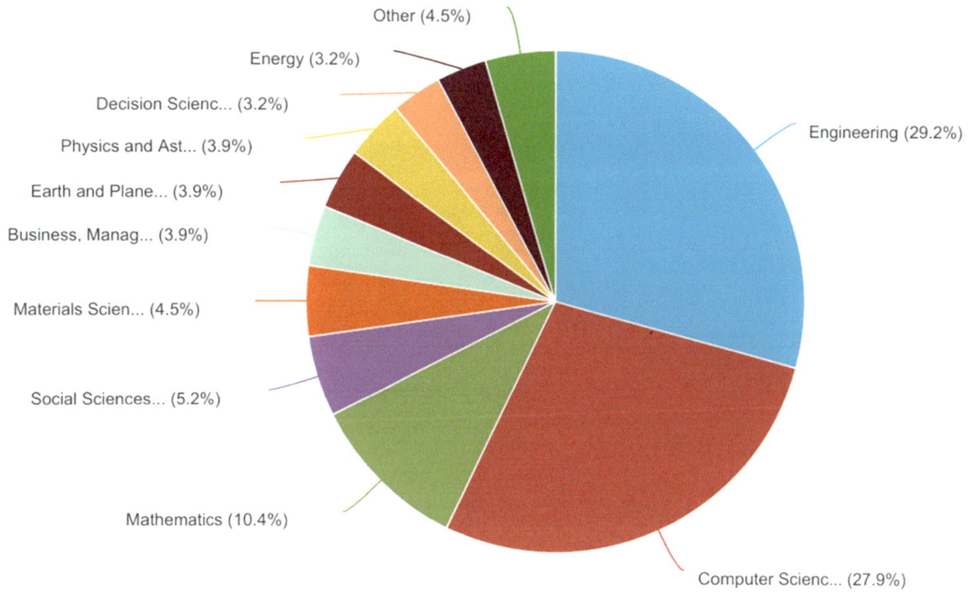

Figure 8.
Location–allocation problem using city HSR vs. field.

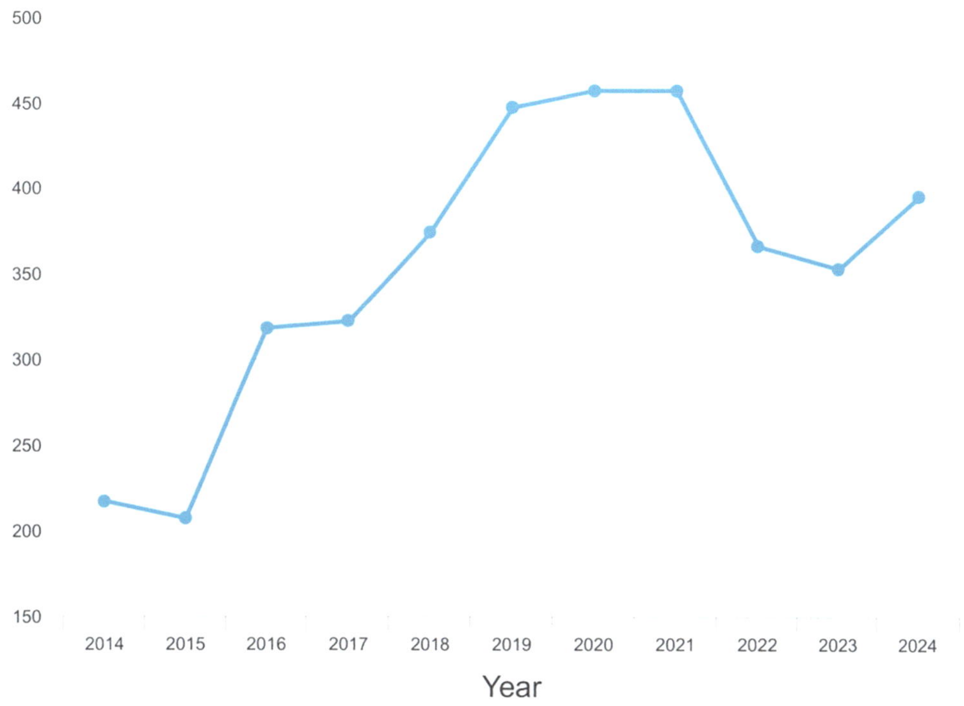

Figure 9.
Publications contributed by researchers.

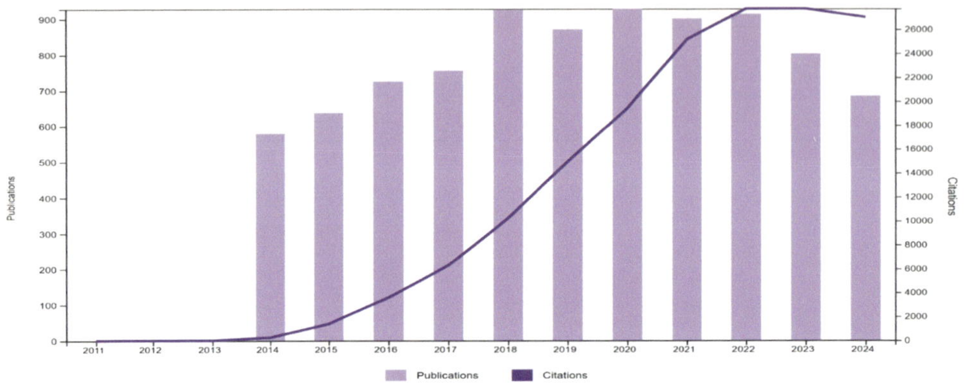

Figure 10.
Publications vs. the year 2014–2024.

"Montreal" + "HSR" to search for articles in the fields of society, engineering, computer science, energy, mathematics, and business over the past ten years, more people have started to study this problem. Researchers approach this topic from different viewpoints and categories, including sustainable development, transportation science and technology, computer science, environmental science, mathematics, business, management, transactions on vehicular technology, and relevant interdisciplinary fields. Most researchers use different algorithms, often metaheuristics in mathematics or applied mathematics. With the development of AI science and technology, an increasing number of researchers tend to apply AI or a combination of AI and metaheuristic algorithms to the location–allocation problem for city logistics using the HSR system. They also consider various constraints such as environmental science, geographical limitations, clean electricity resource demands, population immigration trends, differences between rural and suburban regions, economic development, international transportation, and sustainable demands from customers. In new regions such as the GMA area, these constraints are often key to solving the location–allocation problem for city logistics using the HSR system. Some of those research studies are listed below.

For the location–allocation of urban logistics integrating HSR in the GMA, existing research remains limited. However, over the past five years, a growing body of studies has emerged, evident from keyword searches combining "Montreal" and "HSR" across disciplines such as social sciences, engineering, computer science, energy, mathematics, and business over the last decade. Researchers have approached this topic from diverse perspectives, spanning sustainable development, transportation science and technology, computer science, environmental science, mathematics, business management, vehicular technology, and interdisciplinary fields. Most studies employ methodological frameworks rooted in metaheuristic algorithms within mathematical and applied sciences. With advancements in AI, there is a pronounced trend toward integrating AI – either standalone or in conjunction with metaheuristics – into HSR-based urban logistics location–allocation models. This integration addresses multifaceted constraints, including environmental impacts, geographic limitations, clean energy requirements, population migration patterns, rural–suburban disparities, economic development dynamics, international transportation networks, and evolving customer demands for sustainability.

Articles	Year	Subject area		Source title		Affiliation		Country		Source type	
709	2024	Social sciences	1043	Sustainability Switzerland	44	Beijing Jiaotong University	191	China	747	Journal	1730
358	2023	Engineering	1032	Geriatrics and Gerontology International	42	Université McGill	134	Canada	634	Conference proceeding	335
114	2022	Computer Science	547	International Journal of Molecular Sciences	34	Concordia University	96	United States	306	Book series	51
71	2021	Environmental science	343	Journal of Transport Geography	31	Southwest Jiaotong University	91	United Kingdom	128	Book	29
287	2020	Medicine	278	Archives of Gerontology and Geriatrics	29	University of Montreal	89	Spain	74	Trade journal	21
	2019	Energy	225	Transportation Research Part A: Policy and Practice	29	Université du Québec à Montréal	72	France	68		
	2018	Mathematics	182	IEEE Transactions on Neural Systems and Rehabilitation Engineering	26	Southeast University	55	Australia	62		
	2017	Arts and humanities	167	IEEE Transactions on Vehicular Technology	24	Institut National de la Recherche Scientifique	51	South Korea	62		
	2016	Business, management and Accounting	138	IEEE Transactions on Geoscience and Remote Sensing	23	Centre Urbanization Culture Société	51	India	56		
	2015	Earth and planetary sciences	136	Urban History Review	23	Ministry of Education of the People's Republic of China	40	Germany	47		
	2014	Biochemistry, genetics, and molecular biology	129	Transportation Research Part D: Transport and Environment	22	Polytechnique Montréal	40	Hong Kong	47		

Table 4.
City logistics using the HSR in the GMA.

In summary, understanding these contextual constraints is pivotal for solving location–allocation problems in new regions such as the GMA, where HSR-driven urban logistics models must balance technical feasibility with socioeconomic and environmental imperatives.

Table 4 reveals strategic trends in logistics research (2014–2024) in the location–allocation field across various disciplines by different institutes. Chinese institutions, such as Beijing Jiaotong University and Southwest Jiaotong University, lead in social sciences and environmental science, aligning with China's national logistics strategy and its $1.2 trillion market. McGill University (2023) and Concordia University (2022) exemplify transdisciplinary innovation in engineering and computer science, integrating gerontology with AI-driven optimization.

Publications span journals (e.g., Sustainability Switzerland), conference proceedings, and books, reflecting diverse dissemination channels. Notable is the 2021 Journal of Transport Geography study highlighting UK–China collaboration in sustainable transportation, mirroring global green tech partnerships. Gaps persist in Earth Sciences integration (terrain impact on HSR) and Global South representation, while emerging fields such as electric vertical take-off and landing logistics remain underexplored. The corpus underscores a shift toward AI-empowered, multi-objective frameworks balancing cost, sustainability, and urban resilience.

4. Current research on sustainable location–allocation for urban logistics using HSR in the GMA

The available research on sustainable location–allocation models for urban logistics integrating HSR in the GMA remains limited. To date, only the Canadian government and a handful of scholars have explored HSR system planning and implementation in this region. By contrast, China – with its mature and expanding HSR networks – has emerged as a global leader in this field, with academic studies demonstrating significant depth and innovation. The disparity between Canadian and Chinese research is stark: While the Canadian government proposes a 200-kilometer HSR line between Toronto and Quebec City, its design speed would be categorized as conventional rail in China (where HSR typically operates at ≥250 km/h).

5. Conclusion

This review focuses on sustainable advanced location–allocation for urban logistics leveraging HSR services, examining the feasibility of integrating HSR into logistics networks. Fundamental methodologies rely on metaheuristic algorithms (e.g., GAs, Tabu Search, ACO, SA), with recent advancements incorporating NNs combined with metaheuristics. AI-driven training models for HSR-based urban logistics have also gained traction. Key subproblems – location analysis, allocation strategies, and modeling frameworks – are reviewed across Japan, Europe, and China. Researchers have applied metaheuristics and AI technologies (e.g., ArcGIS, mixed-integer programming, CPLEX optimization) to develop methodologies such as decision tree models, multi-layer perceptron networks, and self-organizing map NNs, successfully addressing urban logistics challenges. Notably, this review identifies a research gap in the GMA and proposes future investigations into sustainable advanced location–allocation models tailored to the region, aiming to inform HSR-integrated urban logistics planning and optimization.

Author details

Yong Lin Ren* and Anjali Awasthi
Concordia University, Montreal, Canada

*Address all correspondence to: yo_re@live.concordia.ca

References

[1] Gogna A, Tayal A. Metaheuristics: Review and application. Journal of Experimental & Theoretical Artificial Intelligence. 2013;**25**(4):5035–5046

[2] Pan W-T. A new fruit fly optimization algorithm: Taking the financial distress model as an example. Knowledge-Based Systems. 2012;**26**: 69–74. DOI: 10.1016/j.knosys.2011.07.001

[3] Bonabeau E, Dorigo M, Theraulaz G. Swarm Intelligence: From Natural to Artificial Systems. Oxford University Press; 1999

[4] Good N, Brown A (2006). Multi-objective concept design of an advanced location allocation delivery system ship (ALDV). In ASNE Joint Sea Basing Symposium. Arlington, VA

[5] Krishnanand KN, Ghose D (2005). Detection of multiple source locations using a glowworm metaphor with applications to collective robotics. In IEEE Swarm Intelligence Symposium (pp. 84–91)

[6] Shi Y, Eberhart RC (1999). Empirical study of particle swarm optimization. IEEE Evolutionary Computation, 1–7

[7] Corne DW, Knowles J, Oates MJ. The Pareto envelope-based selection algorithm for multiobjective optimization. In Parallel Problem Solving from Nature, Springer, 2000. pp. 839–848

[8] Corne DW, Jerram NR, Knowles J, Oates MJ (2001). PESA-II: Region-based selection in evolutionary multiobjective optimization. In Genetic and evolutionary computing conference (pp. 283–290)

[9] Hruschka ER, Campello RJGB, Freitas AA, De Carvalho ACPLF.A survey of evolutionary algorithms for clustering. IEEE Transactions on Systems, Man, and Cybernetics, Part C (Applications and Reviews). 2009;**39**(2):133–155

[10] Biswas A, Dasgupta S, Das S, Abraham A. Synergy of PSO and bacterial foraging optimization—A comparative study on numerical benchmarks. In Innovations in Hybrid Intelligent Systems, Springer, 2007. pp. 255–263

[11] Karaboga D, Basturk B. A powerful and efficient algorithm for numerical function optimization: Artificial bee colony (ABC) algorithm. Computational Optimization and Applications. 2007;**39**(3):171–459

[12] Karaboga D, Basturk B. On the performance of artificial bee colony (ABC) algorithm. Applied Soft Computing. 2008;**8**(1):687–697

[13] Holland JH. Adaptation in Natural and Artificial Systems. University of Michigan Press; 1975

[14] Reeves C. Genetic algorithms. Handbook of Metaheuristics. 2003;**57**:55–82

[15] Ali MM, Storey C. Aspiration based simulated annealing algorithm. Journal of Global Optimization. 1997;**11**(2): 181–191

[16] Glover F. Tabu search, Part I. ORSA Journal on Computing. 1990;**2** (1):2–30

[17] Bilbro GL, Snyder WE. Optimization of functions with many minima. IEEE Transactions on Systems, Man, and Cybernetics. 1991;**21**(4): 840–849

[18] Subramanyam A, Gounaris CE. A decomposition algorithm for the consistent traveling salesman problem with vehicle idling. Transportation Science. 2018;**52**(2):386–401

[19] Marco D. Ant Colony Optimization (ACO) Applied to Vehicle Routing Problems. Advanced Location allocation Systems; 2002

[20] Revelle C, Marianov V. Location analysis. In Encyclopedia of Operations Research and Management Science. Boston, MA: Springer; 2013. p. 892–899

[21] Gambardella LM, Rizzoli AE, Oliverio F, Casagrande N, Donati AV, Montemanni R, Lucibello E (2003). Ant colony optimization for vehicle routing in advanced location allocation systems. In Proceedings of the International Workshop on Modelling and Applied Simulation (MAS 2003)

[22] Kansou A, Kanso B. Hybrid metaheuristics for the periodic open arc routing problem. International Journal of Metaheuristics. 2022;**8**(1):27–50

[23] Zhang Y, Mei Y, Tang K, Jiang K. Memetic algorithm with route decomposing for periodic capacitated arc routing problem. Applied Soft Computing. 2017;**52**:1130–1142

[24] Specht DF. Probabilistic neural networks. Neural Networks. 1990; **3**(1):109–118

[25] Nam HS, Song DW. Defining maritime logistics hub and its implication for container port. Maritime Policy & Management. 2011;**38**(3): 269–292

[26] Miyashita K. Structural change in the international advanced location allocation. The Asian Journal of Shipping and Location Allocation. 2009;**25**(1):121–138

[27] Murakami H. Market performance of low-cost entry into the airline industry: A case of two major Japanese markets. The Asian Journal of Shipping and Location Allocation. 2009;**25** (1):103–120

[28] Chung T-W.A study on selections of strategic type of business in air-location allocation industry clusters. The Asian Journal of Shipping and Location Allocation. 2009;**25**(1):83–102

[29] Purkayastha D, Majumder M, Chakrabarti S. Collection and recycle bin location allocation problem in solid waste disposal. Polymer Journal. 2014;**1**:175–191

[30] Brinckerhoff P, Douglas. (2003). Boston to Montreal high-speed rail planning and feasibility study phase I (pp. 1–3). Available from: https://rosap.ntl.bts.gov/view/dot/57470

[31] Lyu Z, Pons D, Palliparampil G, Zhang Y. Optimising urban freight logistics using discrete-event simulation and cluster analysis: A stochastic two-tier hub-and-spoke architecture approach. Smart Cities. 2023;**6**(5): 2347–2366

[32] Zhang B, Dong W, Yao J. The opening of high-speed railways, the improvement of factor allocation efficiency between regions, and the city's environmental quality improvement. International Journal of Environmental Research and Public

Health. 2023;**20**(5):4648. DOI: 10.3390/ijerph20054648

[33] Khosravi S, Bozorgi A, Zahedi-Seresht M. Hub-and-spoke network design considering congestion and flow-based cost function. Applied Sciences. 2024;**14**(15):6416. DOI: 10.3390/app14156416

[34] Liu Q, Chen Y, Hu W, Dong J, Sun B, Cheng H. Underground logistics network design for large-scale municipal solid waste collection: A case study of Nanjing, China. Sustainability. 2023;**15**(23):1–25

[35] Mahfuzur M, Rahman S, Islam R. Integrating GIS sensing for location allocation modeling in emergency evacuation planning. International Journal of Disaster Risk Reduction. 2020;**45**:101567. DOI: 10.1016/j.ijdrr.2020.101567

[36] Tsadiras AK, Papadopoulos CT, O'Kelly MEJ. An artificial neural network based decision support system for solving the buffer allocation problem in reliable production lines. Computers & Industrial Engineering. 2013;**66**:1150–1162

[37] Hsieh K-H, Tien F-C. Self-organizing feature maps for solving location–allocation problems with rectilinear distances. Computers & Operations Research. 2004;**31**(7):1017–1031

[38] Xenitidis K, Ioannou K, Tsantopoulos G, Myronidis D. On the usage of artificial neural networks for the determination of optimal wind farms allocation. Sustainability. 2023;**15**(24):2–24

[39] Zhang W, Liu Y, Liu T, Yang C (2018). Trajectory prediction with recurrent neural networks for predictive resource allocation. In 2018 14th IEEE International Conference on Signal Processing (ICSP) (pp. 634–639). Beijing, China. DOI: 10.1109/ICSP.2018.8652460

[40] Li S, Lang MX, Chen X, Li S, Liu W, Tang W. Logistics hub location for high-speed rail freight transport with road-rail intermodal transport network. PLOS ONE. 2023;**18**(7):e0288333. DOI: 10.1371/journal.pone.0288333

[41] Zhen L, Fan T, Li H, Wang S, Tan Z. An optimization model for express delivery with high-speed railway. Transportation Research Part E: Logistics and Transportation Review. 2023;**176**:2–15

[42] Pašagić Škrinjar J, Rogić K, Stanković R. Location of urban logistic terminals as hub location problem. Transport Logistics Review. 2012;**24**:433–439

[43] Shahparvari S, Nasirian A, Mohammad A, Noori S, Chhetri P. A GIS-LP integrated approach for the logistics hub location problem. Computers & Industrial Engineering. 2020;**146**:2–15

[44] Wang L, Xiao T, Liu S, Zhang W, Yang B, Chen L. Quantification of model uncertainty and variability for landslide displacement prediction based on Monte Carlo simulation. Gondwana Research. 2023;**123**:27–40

[45] Anderluh A, Nolz PC, Hemmelmayr VC, Crainic TG. Multi-objective optimization of a two-echelon vehicle routing problem with vehicle synchronization and 'grey zone' customers arising in urban logistics. European Journal of

Operational Research. 2021;**289**(3):940–958

[46] Feng ZK, Niu WJ, Wang WC, Zhou JZ, Cheng CT. A mixed integer linear programming model for unit commitment of thermal plants with peak shaving operation aspect in regional power grid lack of flexible hydropower energy. Energy. 2019;**175**:618–629

[47] Li H, Zhang B, Ge X. Modeling emergency logistics Location-allocation problem with uncertain parameters. Systems. 2022;**10**(2):51

[48] Chu H, Zhang W, Bai P, Chen Y. Data-driven optimization for last-mile delivery. Complex & Intelligent Systems. 2023;**9**(3):2271–2284

[49] Ortega E, Monzón A, López E. The influence of spatial data allocation procedures on accessibility results: The case of high-speed rail networks. Applied Geography. 2018;**94**:241–250

[50] Li S, Chen X, Lang M, Li S, Yu X, Geng Y. Optimization of high-speed railway line planning with passenger and freight transport coordination. IEEE Access. 2022;**10**:100340–100353

[51] Repolho HM, Church RL, Antunes AP. Optimizing station location and fleet composition for a high-speed rail line. Transportation Research Part E: Logistics and Transportation Review. 2016;**93**:437–452

[52] Loo BPY, Zhiran. (2023). Statistical analysis of high-speed rail station development in 533 Chinese and European cities (1976–2021)

[53] Wang Y, Lo H. The impact of rail transit on urban spatial structure: A case study of Guangzhou, China. Habitat International. 2016;**59**:1–10

[54] Huang K, Wu TY. A HEURISTIC BASED ON MODIFIED LAGRANGIAN RALAXATION FOR THE VEHICLE ROUTING PROBLEM. Journal of the Eastern Asia Society for Transportation Studies. 2007;**7**:1056–1069

[55] Chu H, Chen Y. A novel hybrid algorithm for multiobjective location allocation problem in emergency logistics. Computational Intelligence and Neuroscience. 2021;**2021**:1951161. DOI: 10.1155/2021/1951161

[56] Ishfaq R, Sox CR. Hub location–allocation in intermodal logistic networks. European Journal of Operational Research. 2011;**210**(2):213–230

[57] Loo BPY, Huang Z. Location matters: High-speed railway (HSR) stations in city evolution. Cities. 2023;**135-139**:1–12